Perspectives on Multimedia

Perspectives on Multimedia

Communication, Media and Information Technology

Robert Burnett, Anna Brunstrom,
Anders G. Nilsson
Karlstad University, Sweden

John Wiley & Sons, Ltd

Copyright © 2003   John Wiley & Sons Ltd, The Atrium, Southern Gate, Chichester,
West Sussex PO19 8SQ, England

Telephone   (+44) 1243 779777

Email (for orders and customer service enquiries): cs-books@wiley.co.uk
Visit our Home Page on www.wileyeurope.com or www.wiley.com

This publication is designed to provide accurate and authoritative information in regard to
the subject matter covered. It is sold on the understanding that the Publisher is not engaged
in rendering professional services. If professional advice or other expert assistance is
required, the services of a competent professional should be sought.

Other Wiley Editorial Offices

John Wiley & Sons Inc., 111 River Street, Hoboken, NJ 07030, USA

Jossey-Bass, 989 Market Street, San Francisco, CA 94103-1741, USA

Wiley-VCH Verlag GmbH, Boschstr. 12, D-69469 Weinheim, Germany

John Wiley & Sons Australia Ltd, 33 Park Road, Milton, Queensland 4064, Australia

John Wiley & Sons (Asia) Pte Ltd, 2 Clementi Loop #02-01, Jin Xing Distripark, Singapore 129809

John Wiley & Sons Canada Ltd, 22 Worcester Road, Etobicoke, Ontario, Canada M9W 1L1

British Library Cataloguing in Publication Data

A catalogue record for this book is available from the British Library

ISBN 0-470-86863-5

Typeset in 11/13pt Times by TechBooks, New Delhi, India
Printed and bound in Great Britain by TJ International, Padstow, Cornwall
This book is printed on acid-free paper responsibly manufactured from sustainable forestry
in which at least two trees are planted for each one used for paper production.

Contents

Foreword

Since the Nineteenth century a feature of most academic and intellectual inquiries has been the progressive move to specialization of knowledge. Research rarely crosses the divides of disciplines. When it does, it is both provocative and extremely valued. In recent years, this specialization of knowledge has been challenged by emerging areas of inquiry, like multimedia, which cross over the divides between scientific inquiry, social science and the arts. In order to understand the emerging new media culture with all its technological convergences, it is imperative that there is a similar convergence in its study. There needs to be dialogue between computer science and its application and use in the multifarious environments and settings in order to make sense of these new technocultural forms.

The initiatives that have led to the publication of *Perspectives on Multimedia* have fostered this kind of dialogue. In effect, the various authors have approached the study of multimedia in a crystalline fashion by tackling its complexity from particular vantage points. Each article stands alone in its area of inquiry, but what makes the book particularly valuable is the interface between articles. From cultural approaches to understanding software to business integration/application of multimedia, from isolating the key distinctions of what defines multimedia to the development of coded architecture of multimedia from image construction and delivery to the place of security, this book is a testament to the necessity for research in multimedia to be multidisciplinary. In effect, it has brought together the diversity of how multimedia is conceptualized in these different settings and fostered a more comprehensive, integrated and unique reading of it as a cultural form that, despite its apparent convergent structure, has dispersed applications and uses. The integrated research climate at Karlstad University that has led to this publication on multimedia will no doubt serve as a model and a source for further inquiry into what has been called our information industries—and in other circles our converged creative industries—that rely so heavily on the integration of technology and culture.

P. David Marshall
Boston, Massachusetts
October 2003

Preface

INTRODUCTION

The term 'multimedia' has become something of a catch-all phrase for a large array of new communication technologies and qualifies without doubt as one of the buzzwords of the past decade, along with similarly fashionable phrases like 'cyberspace' and 'virtual reality'.

A wealth of literature is rapidly accumulating on multimedia developments, particularly in popular and 'how to' genres, and on technical aspects. Critical examinations of multimedia are poorly represented. *Perspectives on Multimedia* is meant to be a contribution to this category of literature and is a collection of reflections on multimedia: its technical and theoretical bases, some of its educational and informational applications, and research approaches and considerations.

The uses of multimedia are rapidly increasing. Its power to present information in ways not previously possible, and its integration of resources, allows for the creation of rich learning environments. In the past, a teacher or student might have to consult many sources and use several media to access the needed information. By integrating media and utilizing hypermedia we are able to create user controlled, information on-demand learning and play environments. Problems occur because of poorly designed programs, resistance to change within organizations, and the lack of technology on which to run multimedia software. As technology costs come down, designers become better at producing programs, and users become more familiar with using multimedia, we will see a growing acceptance of it in all settings. Current uses of multimedia include, but are not limited to: computer-based learning; reference systems; simulations; games; virtual environments; and 'edutainment'.

In the academic arena, multimedia has found a home in the Special Interest Group MultiMedia (SIGMM) of the Association for Computing Machinery (ACM), which provides a forum for researchers, engineers, and practitioners in all aspects of multimedia computing, communication, storage, and applications. SIGMM divides the field of multimedia into three areas: systems, media, and applications.

Systems research includes network protocols, operating system support for digital audio and video, middleware toolkits, streaming media servers, databases, sensors and actuators, and client/server systems. Media research includes processing, coding, and analysis. Specific research topics include media synchronization, content analysis, content-based retrieval, security, asset management, audio/image/video processing, and compression. Applications research includes end-systems, distributed collaboration, and content capture and authoring. Specific research includes hypermedia systems, multimodal interaction, webcasting, distance learning, and distributed virtual environments.

The research program *Communication: Media and Information Technology* (CMIT) at Karlstad University focuses on the space or interface between traditional subject areas. CMIT is one of a limited number of strategic research profiles developed at Karlstad University in cooperation with the Swedish Knowledge Foundation. The CMIT research platform builds upon social scientific, humanistic, and technical competence. The differences between these traditions are gradually being broken down. Many research tasks require interdisciplinary reflection. The convergence of media, information and communication technology into 'multimedia' is an example of an area that needs interdisciplinary reflection. The World Wide Web, electronic newspapers or Web TV are best not studied in isolation but from a multiple of positions and scientific backgrounds.

The aim of CMIT research has been to develop new competence and generate knowledge through interdisciplinary research into the converging vectors of communication, media and information technology. We have examined where this convergence is most visible; in the development and transformation of media, information and communication technologies into hybrid 'multimedia'. This book is one of the many results of this research.

For the purposes of this book, we have divided the contributions into our three main subject areas: media and communication, information systems, and computer science. They represent three perspectives or ways of looking on the field of multimedia. A short overview of the three perspectives and each of the chapters included in the book is provided below.

MEDIA AND COMMUNICATION PERSPECTIVES ON MULTIMEDIA

Media and communication research has developed into a tremendously expansive, dynamic, and multifaceted field. This has broken down the dominance of direct or indirect policy-oriented media research, which was interested in specific instrumental knowledge. The new basic, fundamental, media and communication research is contributing to a deeper understanding of social tendencies and processes in general. The Karlstad media and communication research agenda aims to combine theories of the social, psychological, and aesthetic aspects of new media developed in close, empirical studies from the past or present.

A rapid influx of new technological innovations has caused the media to begin to overlap in their functions and their forms. Much of this communication technology convergence is bringing more control to the user, or, to put it another way, making communication more interactive. As the distinctions between the various communication media begin to blur, the study of the entire communication process becomes important. All communication deals with symbols, regardless of whether it is in the form of a newspaper article, a television program, a radio talk show or a computer program. A central effort is therefore being directed towards the development, application and analysis of media and communication technology in the light of historical, symbolic, societal, political, pedagogical and technological factors. This applies to both the conditions for and the consequences of such technology. The three contributions to this volume on multimedia from a media and communication perspective reveal some of the breadth of this new and exciting subject area.

Chapter 1, by Robert Burnett, looks at the current situation. Many critics of today's multimedia shy away from attempts to define or identify a dominant theme behind the emergence of this new medium. They say that the subject is too diverse, that it resists a neat historical frame. In fact, there is a tendency among critics to celebrate the elusive nature of the subject. There are of course many differing definitions and potential readings of multimedia's history. In this chapter, the key concepts intrinsic to digital forms of multimedia are defined as: integration, interactivity, hypermedia, immersion, narrativity and hybridity. These six characteristics determine the scope of multimedia's capabilities for expression; they establish its full potential. We follow these definitions to see how their characteristics evolved more or less simultaneously, each following its own tradition and yet inextricably interwoven with the others in a web of mutual influence.

Chapter 2, by Steve Gibson, discusses the interface between humans and machines, specifically some alternative approaches to human interaction with large-scale digital systems. The chapter focuses on those interface strategies that purposely diverge from the standard mouse–keyboard arrangement and which, therefore, offer a wider range of sensory interaction with digital systems. The central focus of this discussion will revolve around the collaborative multimedia performances *Cut to the Chase* and *telebody*, though we will also refer to other projects currently engaged in alternative interface research and development. It should be noted that the present discussion is by no means exhaustive. The choices we have made are highly personal and represent our on-going interest not only in interface technology, but also in the philosophies surrounding new technology and cyberculture. For the sake of comparison we have chosen approaches that are in some cases diametrically opposed to each other.

Chapter 3, by Andreas Kitzmann, examines the narratives of emancipation and empowerment that often form the rhetorical background to software development and application. In particular, it focuses on a prototype multimedia authoring program called *Night Kitchen* which is designed to provide the common user with fuller access to the powers of multimedia programming and design. Of particular interest is the notion that the *Night Kitchen* project is partially undermined by the standardization that is said inevitably to accompany the design of so-called 'user friendly' software. In the course of exploring such an issue, this chapter works towards the conclusion that software is not only a tool but also an important aspect of cultural expression, and thus worthy of serious critique and analysis. Such a project, however, is not without its challenges, especially with respect to the question of how much technical knowledge is necessary to make such a task both possible and meaningful.

INFORMATION SYSTEMS PERSPECTIVES ON MULTIMEDIA

Information systems science is an expansive area in our present society. It is the scientific discipline that studies human interaction with information systems (IS) in different business settings. The subject focuses on developing knowledge of the use of information technology (IT) in society and in the business community. In this knowledge creation there are always questions as to how using modern information technology enables us to help private companies and public organizations to support business operations, create new business opportunities, and promote a more proactive service management.

Finding the answers to these questions is what we consider to be the primary mission of information systems science. Developing business operations should be regarded as a knowledge promotion, where technology is subject to the needs of human beings, i.e. is based on a user-oriented perspective. The reason for this is to place information technology in a greater context—people and organizations should become better at creating and using IS/IT solutions in our society.

Research in the subject information systems involves, for example, issues on systems development methodologies, use of enterprise systems (ERP-systems) in organizations and requirements specifications for change work. Studies in multimedia have become a new area of interest or challenge for the information systems discipline. Multimedia research in information systems is today focused on models and methods for development and evaluation of interactive media solutions in organizations. We will demonstrate this with four contributions illuminating issues on business modelling for multimedia development, requirements engineering of interactive media, evaluation of interactive multimedia, and conceptual modelling of multimedia databases.

As detailed in Chapter 4, by Anders G. Nilsson, by business modelling we refer to how different actors or people are using models and methods in order to understand and change business processes together with their information systems in our organizations. From earlier research it is a well-known fact that there are real communication gaps between top managers, business leaders and systems experts when developing organizations. In this context we can distinguish three levels of development work, namely strategy development, process development and systems development. The purpose of business modelling is to bridge the communication gap between actors and create a good coordination between the three development levels in organizations. Development of multimedia products is a rather new form of change work in today's organizations. Multimedia development can affect key issues on all three development levels and it is therefore interesting to study the role of digital media when developing future organizations.

Chapter 5, by Lennart Molin and John Sören Pettersson, deals with the specifying of requirements for a multimedia system, which is difficult and often carried out in an informal way. Requirements engineering in traditional information systems development is, on the other hand, often done in a more formal and thorough way. A large proportion of the functionality has to be defined relative to end-users who are not experts in formulating system requirements. One way to identify and formulate the requirements would be to use a tool permitting demonstrations and tests of ideas regarding interactive products. This is the motive for an ongoing construction of an experimental station, Ozlab, for interactive user interfaces. Ozlab permits mock-up prototyping where the prototype looks real but is directly controlled by a test leader. The set-up is intended to admit even non-programmers to test ideas for interactive products, and this survey ends up with a suggestion for a requirements gathering procedure for laymen based on the Ozlab system.

Chapter 6, by Louise Ulfhake, deals with evaluation, which is important to assure quality during design and development of interactive multimedia, as well as in reviewing of the final product. This chapter presents an evaluation method, which is both inexpensive and timesaving. Six European business schools collaborated in the years 1998–2000 (Esprit

IV program) in the production of several multimedia cases on small and medium sized companies as well as on large enterprises. One of the business case studies is used to illustrate more thoroughly and exemplify a constructive evaluation method. In critically evaluating the multimedia cases, it is found that systematic models and methods have a great deal to offer in the field of quality interactive multimedia. This survey proposes an evaluation method, consisting of four integrated parts or steps: (1) evaluation of the structure, (2) evaluation of the interaction, (3) evaluation of the usability, and finally (4) evaluation of the productivity.

The purpose of the survey in Chapter 7, by Lars Erik Axelsson, is to expose and explain some problems on conceptual modelling in multimedia databases. By way of introduction, a description is given of the idea of conceptual modelling starting from the traditional ISO report where some fundamental concepts and definitions are described. The modelling techniques and their capability or limitations concerning modelling of business environments in order to create multimedia databases are discussed. The discussion indicates that the general concepts and definitions in conceptual modelling suggested in the ISO report comprise the foundation for analysing more complex information, such as in multimedia contexts. The survey also proposes that multimedia characteristics (e.g. interactivity and narrativity) and emotional factors on communication and information are key factors in the process of understanding how conceptual modelling can support multimedia databases.

COMPUTER SCIENCE PERSPECTIVES ON MULTIMEDIA

Computer science provides the technical foundation for the computer software and hardware that drives the information age. It is a discipline that involves the understanding and design of computational processes and structures. The discipline spans from theoretical studies of algorithms to practical problems of software and hardware implementation. The integration of theoretical studies, experimental methods, and engineering design into one discipline sets computer science apart from many other areas. Although the importance of experimentation within computer science may be partly related to the relative youth of the field, it is primarily due to the complexity of the studied artifacts. The importance of experimentation is reflected in this book in that several of the computer science chapters support the presented ideas with experimental results.

Computer science is a synthetic field in that the object of study is not a natural phenomenon. Computer science studies software and hardware artefacts created by humans. The creation and demonstration of new computational artefacts is an important part of the discipline. The focus on artefacts creates a highly dynamic field, restricted only by the limits of our imagination. It also explains the continuing emergence and evolvement of new application areas for computer science. In the last decade, multimedia has become one such very important application area.

When studying multimedia from a technical or computer science perspective many interesting and important problems emerge. Key research issues include how multimedia can best be represented for various purposes, how multimedia is stored and retrieved and how multimedia is transferred over a network, as well as issues related to the presentation of multimedia. Keeping with the focus of the book we will concentrate on issues related to the communication of multimedia. In particular, the four computer science chapters of the book all relate to the transfer of multimedia over the Internet.

Chapter 8, by Stefan Lindskog and Erland Jonsson, covers QoS and security. In a Quality of Service (QoS) aware communication system, a user is able to choose between various service classes, each with different reliability, predictability, and efficiency degrees. However, until now security has not been recognized as a parameter in QoS architectures and no security-related service classes have been defined. This implies that end-users have no chance of configuring their level of security, which is remarkable. This chapter contains a survey of QoS architectures as seen from a security point of view and presents some initial ideas on how QoS architectures can be extended with a security dimension.

Chapter 9, by Katarina Asplund and Anna Brunstrom, introduces the idea of a partially reliable service. The ongoing deployment of applications that transmit multimedia data makes it important for the Internet to better accommodate the service requirements of this type of application. One approach is to provide a partially reliable transport service, i.e. a service that does not insist on recovering all, but just some of the packet losses, thus providing lower delay than a reliable transport service. This chapter describes a transport protocol that provides such a partially reliable service. The protocol, which is called PRTP, is especially aimed at applications with soft real-time requirements. Experimental results that illustrate the potential performance gain of partial reliability are presented.

Multimedia usage among consumers is an increasing market. Another trend is toward increasing usage of mobility and wireless freedom. Chapter 10, by Stefan Alfredsson and Anna Brunstrom, explores the possibilities of increasing communication between layers, in order to improve network and application performance. Multimedia applications may not need the full reliability provided by the transport layer, but would prefer, for example, faster transmission. Wireless networks can have high error rates that degrade the transmission rate when full reliability has to be maintained. By accepting these errors, the transmission rate can be improved, provided the application can handle the residual errors in the data. How the transport layer can be extended with this functionality is presented and the resulting performance implications explored.

Delivery of multimedia content can be done over a wide variety of communications technologies and to end-user devices with different capabilities. Chapter 11, by Johan Garcia and Anna Brunstrom, discusses transcoding, which is one way to address the heterogeneity in different communication and device technologies. Transcoding can be applied to any digital multimedia data, and in this chapter transcoding of web images is used to illustrate some of the issues surrounding transcoding. It is shown both how JPEG images can be efficiently transcoded to reach higher compression ratios as well as how they can be transcoded to a more robust format.

R.B.
A.B.
A.G.N.

1

Multimedia: Back To The Future!

Robert Burnett

Media and Communication, Karlstad University

1.1 INTRODUCTION

It is easy to argue that digital technology ever increasingly mediates our relationship to the world, whether it be through the telephone, television, computer, or the Internet. *Multimedia* has, since the 1990s, taken it's place in the consumer commodity marketplace, one of the most recent in successive waves of new media that have arrived upon us with much hyperbole.

Multimedia has not been feasible for the home consumer until recently because computers had not been able to deliver an integrated package at an affordable price. In 1975 the first personal computers were marketed with low processor power, black and green text-only screens, and were used for accounting and inventory. By 1980 we saw the addition of hard disk storage and simple graphics for forecasting and statistics. By around 1985 we saw the capability of displaying colour, more advanced

Perspectives on Multimedia R. Burnett, Anna Brunstrom and Anders G. Nilsson
© 2004 John Wiley & Sons, Ltd ISBN: 0-470-86863-5

graphics, sounds, and animation for word processing and desktop publishing.

Since about 1995 we have had the capability of integrating digital video, sounds, animation and text into one hardware and software package. There is increasing emphasis on communications capabilities and sharing information over networks such as the Internet. Multimedia is made possible and affordable today because of increases in storage and speed and decreases in size and cost; this yields an increase in performance and availability.

Of course computers have ubiquitously overtaken our desktops, offices and homes, and move data to and from these places. Computers both help to produce our multimedia experience and allow us access to it. The computer is a language machine because it is programmable— as such, operating symbolically upon symbolic things. To paraphrase Turing, the computer is a medium that can be any medium. From this definition it is easy to see the rapid technical development of the computer as the very means for what we now call multimedia.

The most common technological descriptions of multimedia are usually something like, 'the integration of two or more different media with the personal computer'. The component media including text, graphics, animation, speech, music and video. Not many years ago, multimedia meant multiple slide projectors and a tape deck. Take for example the definition by Burger (1993) in the *Desktop Multimedia Bible*: 'Multimedia is the combination of two or more media'; or in the *Merriam-Webster's Collegiate Dictionary* (1996): 'Using, involving, or encompassing several media'.

Compare these past definitions with the recent *Encyclopædia Britannica* (2001) entry:

> Any computer-delivered electronic system that allows the user to control, combine, and manipulate different types of media, such as text, sound, video, computer graphics, and animation. Interactive multimedia integrate computer, memory storage, digital (binary) data, telephone, television, and other information technologies. Their most common applications include training programs, video games, electronic encyclopaedias, and travel guides. Interactive multimedia shift the user's role from observer to participant and are considered the next generation of electronic information systems.

After looking at different definitions we should consider the unique characteristics that make up multimedia. Three properties of multimedia are delineated by van Dijk (1999) as *stratification, modularity,* and *manipulability* of information. Stratification refers to the fact that users can find more information about a fact retrieved in the shape of explanations, figures, illustrations, photos, videos, animations, sounds, and so forth. The same information can be portrayed in several ways. Modularity is based on the fact that an information database is composed of pieces of information to be retrieved separately and combined in whatever way the user wants. The third characteristic is the manipulability of information in multimedia, enabling the user to reassemble pieces of information.

Perhaps the most useful way of defining multimedia is as the links between several devices in one interactive medium, or links between several media in one interactive device. Various applications with sound, text, data, and images can be integrated in a combination of several devices or in a single device. The main characteristics of multimedia are thus the integration of several types of data and high level interactivity caused by control the user has over the interaction.

1.2   MULTIMEDIA AS ART AND SCIENCE

In many discussions of multimedia, the argument has focused upon a dialectic between the arts and sciences—a recapitulation of C.P. Snow's much dated dualism between the 'two cultures'. For heuristic purposes one can argue that the basis of multimedia is based on two differing sciences: communication science and computer science. The definitions and roots of communication science are mostly derived from social sciences and humanities, the so-called 'soft sciences', while the history of computer science is mostly found in the 'hard sciences'. Sometimes this distinction is defined as the difference between Art and Science.

Communication and computer sciences are distinct and yet share many similarities. Some of these include information transmission, usage of electronic media, and information representation (e.g. information theory; minimum code required to represent information without loss). Computer science concentrates on technical aspects of representation, manipulation, transmission, and reception of information, while communication science concentrates on human aspects of the same.

Computer science deals with representation of images in terms of pixels, while communication science deals with representation of content at a theoretical level, more from a presentation point of view. Broadly speaking, computer science deals with transmission of data, while communication science deals with transmission of information/emotion/knowledge. This is the important distinction between *communications* versus *communication*. Communications refers to technical aspect: computer to computer. Communication refers to the human aspect, such as human to human, human to computer, computer to human.

1.3   SECRET HISTORY OF MULTIMEDIA

In their recent book *Multimedia: from Wagner to Virtual Reality*, Packer and Jordan (2001) argue that the concept of integrated, interactive media has its own long history, an evolution that spans over 150 years. Remarkably, this has been a largely untold story. Packer and Jordan present us with a 'secret history' of multimedia: a narrative that includes the pioneering activities of a diverse group of artists, scientists, poets, musicians, and theorists from Richard Wagner to John Cage, from Vannevar Bush to Bill Viola, William Burroughs to William Gibson.

Packer and Jordan (2001) make a strong argument that multimedia, by its very nature, is open, democratic, non-hierarchical, fluid, varied, and inclusive. Multimedia can be defined by five simultaneously working processes:

> Integration, the combining of artistic forms and technology into a hybrid form of expression. Interactivity, the ability of the user to directly manipulate and effect her experience of media. Hypermedia, the linking of seperate media elements to one another to create a trail of personal association. Immersion, the experience of entering into the simulation or suggestion of a three dimensional environment. Narrativity, aesthetic and formal stategies that derive from the above concepts, and which result in non-linear story forms and media presentation.
>
> (Packer and Jordan, 2001, p. xxviii)

Let's look a little closer at these five working components of multimedia and take a brief look at just a handful of the past pioneers who helped pave the way for multimedia as we know it today.

1.3.1 Integration

Integration is the combining of artistic forms and technology into a hybrid form of expression. The first modern attempt of combining art forms can be traced to 1849 when Richard Wagner introduced the concept of the *Gesamtkunstwerk*, or Total Artwork, in an essay called 'The Artwork of the Future.' Wagner's description of the Gesamtkunstwerk is an attempt to establish a practical, theoretical system for the comprehensive integration of the arts. Wagner sought the idealized union of all the arts through the 'totalizing,' or synthesizing, effect of music drama—the unification of music, song, dance, poetry, visual arts, and stagecraft (Packer and Jordan, 2001, p. xviii)

The video artist Nam June Paik's breakthough work in video art in the 1960s developed non-traditional performance techniques that challenged accepted notions of form, categorization, and composition, leading to the emergence of genres such as the Happening, electronic theatre, performance art, and interactive installations. After Paik it was possible to think in terms of 'multimedia' performances and artwork, breaking down the traditional boundaries between differing artforms.

1.3.2 Interactivity

Interactivity (the ability of the user to directly manipulate and effect her experience of media) is the amount of control the user has over the presentation of information. 'Interactive multimedia' refers to multimedia that allows for user control. The three most common classifications of interactivity are: linear, branching, and hypermedia.

An interactive linear presentation is one in which the author decides the sequence and manner in which information is presented. The user controls only the pace. An interactive branching program is one in which the user has some control over the sequence of presentation by selecting from a group of choices such as from a main menu. The author still maintains control of deciding what to include in the choices available at any point in the program. Interactive hypermedia can be thought of as a web of interrelated information in which the user is in almost complete control of the pace, sequence and content of the presentation. Links provide for random access of information.

The musical performance work of John Cage in the 1950s was a significant catalyst in the continuing breakdown of traditional boundaries between artistic disciplines. Together with choreographer Merce Cunningham and artists Robert Rauschenberg and Jasper Johns, Cage devised theatrical experiments that furthered the dissolution of borders between the arts. He was particularly attracted to aesthetic methods that opened the door to greater participation of the audience, especially if these methods encouraged a heightened awareness of subjective experience. Cage's use of indeterminacy and chance-related technique shifted responsibility for the outcome of the work away from the artist, and weakened yet another traditional boundary, the divide between artwork and audience (Packer and Jordan, 2001, p. xviii)

One of the first theoretical attempts to integrate the emerging fields of human–computer interactivity and cybernetics with artistic practice is Roy Ascott's article, 'Behavioral Art and the Cybernetic Vision,' from 1966–67. Ascott noted that the computer was 'the supreme tool that . . . technology has produced. Used in conjunction with synthetic materials it can be expected to open up paths of radical change in art.' Ascott saw that human–computer interaction would profoundly affect aesthetics, leading artists to embrace collaborative and interactive modes of experience (Packer and Jordan, 2001, p. xx)

1.3.3 Hypermedia

Hypermedia covers the linking of separate media elements to one another to create a trail of personal association. Vannevar Bush believed that there must be a better answer to 'how information would be gathered, stored, and accessed in an increasingly information-saturated world' than filing and searching through layers of classification, for as far as the act of combining records is concerned, 'the creative aspect of thinking is concerned only with the selection of the data and the process to be employed and the manipulation thereafter is repetitive in nature and hence a fit matter to be relegated to a machine'(Bush, 2001).

Ted Nelson, who was greatly influenced by Bush's article 'As We May Think', coined the term Hypertext (Nelson, 2001). The hypertext 'exist[s] as part of a much larger system in which the totality might count more than the individual document' (Landow and Delany, 2001) The process of assembling information via hypertext mirrors the structure of the mind by operating by association.

> 'With one item in its grasp, it snaps instantly to the next that is suggested by the association of thoughts, in accordance with some intricate web of trails carried by the cells of the brain. It has other characteristics, of course; trails that are not frequently followed are prone to fade, items are not fully permanent, memory is transitory. Yet the speed of action, the intricacy of trails, the detail of mental pictures, is awe-inspiring beyond all else in nature.'
>
> (Bush, 2001)

Hypertext possesses an almost unlimited power to manipulate texts through its ability constantly to shift meaning by assembling networks of text into new contexts and juxtapositions.

Ted Nelson devised an elaborate system for the sharing of information across computer networks. Called Xanadu, this system would maximize a computer's creative potential. Central to Nelson's approach was the 'hyperlink', a term he coined, inspired by Bush's notion of the Memex's associative trails. Hyperlinks, he proposed, could connect discrete texts in non-linear sequences. Using hyperlinks, Nelson realized, writers could create 'hypertexts,' which he described as 'non-sequential writing' that let the reader make decisions about how the text could be read in other than linear fashion. As he observed in his landmark book from 1974, *Computer Lib/Dream Machines*, 'the structures of ideas are not sequential.' With hypertext, and its multimedia counterpart, 'hypermedia,' writers and artists could create works that encourage the user to leap from one idea to the next in a series of provocative juxtapositions that present alternatives to conventional hierarchies (Packer and Jordan, 2001, p. xxv).

In 1989, Tim Berners-Lee's vision of the World Wide Web, as he designed it, combined the communications language of the Internet with Nelson's hypertext and hypermedia, enabling links between files to extend across a global network. It became possible to link every document, sound file or graphic on the Web in an infinite variety of non-linear paths through the network. And instead of being created by a single author, links could be written by anyone participating in the system. Not only did the open nature of the Web lend itself to a wide array of interactive, multimedia experiences, but by hewing to a non-hierarchical structure and open protocols, Berners-Lee's invention became enormously popular, and led to an explosion in the creation of multimedia.

The success of the Web seemed to confirm the intuition of artists engaging in digital media that in the future, a global media database would inspire new forms of expression (Packer and Jordan, 2001, p. xxviii)

1.3.4 Immersion

Immersion is the experience of entering into the simulation or suggestion of a three dimensional environment. William Gibson's best selling novel *Neuromancer* from 1984 described in detail a future in which virtual reality was a fact of life. Gibson's characters inhabited a virtual environment made possible by the networking of computers, which he named 'cyberspace.' Gibson's cyberspace provided the first literary definition for the computers, hubs, servers, and databases that make up the matrix. His discussion of cyberspace was so enticing, with its suggestion that any computer hacker could 'jack-in to the matrix' with an encounter with an avatar–that it became a touchstone for every engineer, artist and theorist working in the field (Packer and Jordan, 2001, p. xxiii).

To date, different research teams have developed such unique immersive technologies as three-dimensional modeling and animation, video rendering methods from multiple projections, and immersive three-dimensional sound environments. Substantial progress in advanced telepresence, data compression, and wireless communications are being made.

Specifically, immersion is achieved through the use of rich multimodal (visual, auditory, haptic) interface techniques to support the creation and use of multimedia information. High-performance wired and wireless communication, transmission, and compression techniques and mechanisms are used in dissemination and to support multimedia information immediacy and portability. Multimedia database structuring, indexing and management techniques are employed to support information storage, access, sharing and customization.

CMIT has developed the Virtual Reality Light Lab (VRLL) as the test bed framework for the integration of various media. Research projects showcase the interactive use of multimedia information. The VRLL provides a number of key features, including high levels of integration and functionality of its various components. It is uniquely positioned to provide researchers with insight into media integration issues on an ongoing basis (see Gibson, Chapter 2 of this book).

With current technological advances in computer graphics and animation, high-speed networking, signal and image processing, and multimedia information systems, it is now feasible to immerse a person in an immersive information and communication environment. One goal for such immersive environments is for people to interact, communicate and collaborate naturally in a shared virtual space while they reside in distant physical locations. Another goal is to immerse people in information-rich environments that aid in understanding and relating raw data to high-level problems and tasks. Future work includes focusing on different technological and human challenges in realizing such immersive environments.

1.3.5 Narrativity

Narrativity covers the aesthetic and formal stategies that derive from the above concepts, and which result in non-linear story forms and media presentation. William Burroughs was deeply suspicious of established hierarchies. He was especially interested in writing techniques that suggest the spontaneous, moment-by-moment movement of the mind, and how non-linear writing might expand the reader's perception of reality. Through his use of the cut-up and fold-in techniques, which he described in his 1964 essay, 'The Future of the Novel', Burroughs treated the reading experience as one of entering into a multidirectional web of different voices, ideas, perceptions, and periods of time. He saw the cut-up as a tool that let the writer discover previously undetected connections between things, with potentially enlightening and subversive results. With the cut-up, Burroughs prefigured the essential narrative strategy of hypertext and its ability to allow readers to leap across boundaries in time and space (Packer and Jordan, 2001, p. xxvi).

Media artists whose roots lay in performance and video also began investigating hypermedia as a means of exploring new forms for telling stories. Artists such as Bill Viola were drawn to the computer's ability to break down linear narrative structures. Viola approached the medium as a repository for evocative images that could be projected on screens in installations, with the viewer wandering through some three-dimensional, possibly life-sized, field of prerecorded or simulated scenes evolving in time (Packer and Jordan, 2001, p. xxvii).

1.4   MULTIMEDIA AS ART AND PERFORMANCE

In the last few decades, the art world has been flooded with a number of terms invented to define the developing artforms employing the so-called 'new' technologies; for example, multimedia art, electronic arts, digital arts, media arts, new media and cyberarts. All these terms have been variously useful in defining historically specific developments in contemporary artistic practice. For example, the term 'electronic arts' is historically specific to some art practices from the 1960s till the early 1980s, which were based on and operated via electronic systems. The term 'digital arts' is also historically specific to the digitization technologies brought about by developments in computer graphics. These technologies have themselves been superceded by the so-called 'cybertechnologies' of which digitization is merely one aspect.

The most recent term that has been invoked to refer to these technologically driven developments in contemporary art is 'new media'. Manovich, in his recent book, The *Language of New Media* (2001), identified five characteristics that conceptually distinguish 'new media' from previous art forms. These are, namely, *numerical coding*, which facilitates the programmability of the media; *modularity*, which creates a structural discreteness of its parts; *automation* of its production and access; *variability*, meaning that the media can continue to be presented in variable formats and versions well after its 'completion'; and finally, *transcoding*, insofar as its codes operate between and are, therefore, transferable across, different systems.

Metaphorically speaking, the superimposition of 'binary' over 'iconic' code anticipated the convergence of media and computer that followed about 50 years later: 'All existing media are translated into numerical data accessible for the computer. The results: graphics, moving images, sounds, shapes, spaces, and texts become computable, that is, simply sets of computer data. In short media become new media'(Manovich, 2001). The meeting of media and computer, and the computerization of culture as a whole, changes the identity of both media and the computer itself—whereby, as Manovich asserts, 'the identity of media has changed even more dramatically than that of the computer' (Manovich, 2001).

Let's take a closer look at these five components of new media. First, all new media objects are composed of digital code, they are numerical representations. Two key consequences follow from that: new media objects can be described formally, i.e. by using a mathematical function, and they can be subjected to algorithmic manipulation. Media thus become programmable.

Second, all new media objects have a modular structure, i.e. they consist of discrete elements that maintain their independence even when combined into larger objects. A Word document, as well as the World Wide Web, consists of discrete objects which can always be accessed on their own. Modularity thus highlights the 'fundamentally non-hierarchical organization'.

Thirdly, the numerical coding of media and the modular structure of a media object (i.e. the first two principles) allow, according to Manovich, 'for the automation of many operations involved in media creation, manipulation, and access.' Thus, 'human intentionality can be removed from the creative process, at least in part' (Manovich, 2001). Examples for automation can be found in image editing, chat bots, computer games, search engines, software agents, etc.

The fourth principle of new media, deduced from the more basic principles—numerical representation and modularity of information—is variability. New media objects are not 'something fixed once and for all, but something that can exist in different, potentially infinite versions' (Manovich, 2001). Film, for example, whose order of elements is determined once and for all, is diametrically opposed to new media whose order of elements is essentially variable (or 'mutable' and 'liquid'). Examples for variability would be customization and scalability.

The fifth principle, and the 'most substantial consequence of the computerization of media' (Manovitch, 2001), is transcoding. Transcoding basically means translating something into another format. However, the most important aspect is that the structure of computerized media (which, on the surface still may look like media) 'now follows the established conventions of the computer's organization of data' (Manovitch, 2001). Structurewise, new media objects are compatible with, and transcodable into, other computer files. On a more general ('cultural') level, the logic of a computer 'can be expected to influence significantly the traditional cultural logic of media' (Manovitch, 2001); that is, we can expect the 'computer layer' to affect the 'cultural layer'. Manovich (2001) uses the term 'cultural interface' to describe a 'human–computer–culture interface—the ways in which computers present and allow us to interact with cultural data'.

While, Manovich's conceptual clarification of what constitutes 'new media' is incisive and useful for our understanding of many of the contemporary developments in art and technology, there is no reason why the term 'new media' is more appropriate than 'multimedia'. The term 'new' in new media, is conceptually empty insofar as what constitutes the 'new' at any point in time is so variable as to be impossible to identify. The use of the word 'new' also does not facilitate a better theoretical framing or understanding of the peculiar artistic and/or technological developments of the art works. However, given the theoretical value of the above mentioned characteristics to illuminate our understanding of the ongoing developments, it would be useful to coopt them into our understanding of multimedia.

1.5   SUMMING UP

The mathematician Norbert Weiner defined cybernetics as the study and strategic deployment of communicative control processes within complex systems constituted by hierarchically ordered entities. By this he initiated a revolutionary development in the way we have come to think about information and control. Cybernetic systems are thus conceived to be made up of information flows between differently constituted entities like humans, computers, animals and even environments. The flow of information was conceived as a principle explaining how organization occurs across and within multiple hierarchical levels. This meant that seemingly bounded entities could be translated/codified into information, thereby enabling interfaces and easy interactions between them. It is, in fact, arguable that in the last two decades a large amount of technological innovation has been towards greater cyberneticization. This means that in addition to innovations that allow existing technologies to become integrated with each other through cross-platform operability, the 'new' in many 'new technologies' have been exactly their ability to 'hybridize' previously separate functionalities, e.g. web-integrated mobile phones, biochips, artificial life, etc. I see this 'hybridity' as differing from Packer's and Jordan's category 'integration' in enough ways to argue for it's existence as an independent essential sixth category of multimedia. It is this hybridity, the desire to translate, different physical entities and processes into information as well as the control afforded thereby that distinctly characterizes and enables what has come to be called multimedia (Burnett and Marshall, 2003).

Thus, one can sum up that the term multimedia refers to all those art forms, practices and processes that are produced and mediated by the continuing developments in cybertechnologies, specifically in information, communication, imaging, experiential, interface and biotechnologies. Multimedia as defined by contemporary art practice includes the following: digital imaging (whether as digital painting, digital photography and digital video); computer animation; holographic art; virtual reality environments, including gaming; robotic arts; net art, including works in hypertext and telematics; human–machine interfaces (e.g. cyborg technologies); bio-arts that employ biotechnologies (e.g. DNA music, transgenic art, artificial life); computer music and sound arts; and hybrid art works involving interaction with other art forms (e.g. theatre, dance, installations, etc.).

It is disappointing that interactivity in multimedia still tends to involve boring 'point and click' action on the part of the user. This unfortunate elevation of interface over content and meaning is a result of software dominating narrative form. The reason for the adoption of the 'software' domination model is simply that most of us still believe that it is the only, or the best approach to multimedia interactivity.

Now the so-called 'interactive' media have the potential to absolve writers and artists from the illusion of authorial control in much the same way that photography broke the naturalist illusion in art, exposing it not as an inevitable form, but as another set of conventions. Indeed one can contend that authors must attempt to transcend the established syntax of earlier forms of multimedia and invent a coherent artistic language for interaction.

While spatial analogues of narrative remain the dominant form of most game like adventures on CD-ROM, such forms are merely a convention. In virtual reality they are derived from the natural need for a participatory spatial environment. In multimedia all the imagery is pre-created. Uniquely in virtual reality, only the model is generated. Users create their own narrative journey on each engagement (see Chapter 2).

The role of the artist and engineer is challenged in the construction of such immersive narrative environments. The new role begins to resemble the designer of a model and, although the artist/engineer may describe its properties in great detail, she is no longer author of the events set in motion by the user. The participatory aspect of audience/user as performer is also evident in most virtual reality sessions, where participants can create their own stories within the broad boundaries set by the creators.

Multimedia has now a long history as the 'buzzword' of the technovision dream of convergence between computers, media and the Internet. This convergence into multimedia was supposed to facilitate the emergence of a digital driven hypertext on a global scale. Nelson's 'uber' hypertext, Xanadu (an interactive digital system, in which all forms of cultural expression, past, present and future, in all their forms, could be stored and recombined) has not happened, because there is little interest in it and nobody can make it economically feasible. While some of us wait for Xanadu, it would be wise to learn more about the history of multimedia, especially if we want to make the future work.

1.6 REFERENCES

Ascott, R. (2001) 'Behaviourist Art and the Cybernetic Version', in Packer, R. and K. Jordan (Eds) *Multimedia: From Wagner to Virtual Reality,* Norton, New York.

Berners-Lee, T. (1989) *Weaving the Web*, Orion Business Books, London.

Burger, J. (1993) *The Desktop Multimedia Bible*, Addison-Wesley, Reading, Mass.

Burnett, R. and D. P. Marshall (2003) *Web Theory: An Introduction*, Routledge, London.

Burroughs, W. (2001) The Future of the Novel', in Packer, R. and K. Jordan (Eds) *Multimedia: From Wagner to Virtual Reality*, Norton, New York.

Bush, V. (2001) 'As We May Think', in Packer, R. and K. Jordan (Eds) (2001) *Multimedia: From Wagner to Virtual Reality*, Norton, New York.

Gibson, W. (1984) *Neuromancer*, Ace Books, New York.

Landow, G. and P. Delany (2001) 'Hypertext, Hypermedia and Literary Studies: The State of the Art', in Packer, R. and K. Jordan (Eds) *Multimedia: From Wagner to Virtual Reality*, Norton, New York.

Manovich, L. (2001) *The Language of New Media*, MIT Press, Cambridge, Mass.

Nelson, T. (2001) 'Computer Lib/Dream Machines', in Packer, R. and K. Jordan (Eds) (2001) *Multimedia: From Wagner to Virtual Reality*, Norton, New York.

Packer, R. and K. Jordan (Eds) (2001) *Multimedia: From Wagner to Virtual Reality,* Norton, New York.

Van Dijk, J. (1999) *The Network Society*, Sage, London.

Wiener, N. (2001) The Human Use of Beings', in Packer, R. and K. Jordan (Eds) *Multimedia: From Wagner to Virtual Reality,* Norton, New York.

2

Alternative Approaches to Interface Technology

Steve Gibson
University of Victoria

2.1 INTRODUCTION

The central focus of this discussion will revolve around the collaborative multimedia performances *Cut to the Chase* and *telebody*, though I will also refer to other projects currently engaged in alternative interface research and development. It should be noted that the present discussion is by no means exhaustive. The choices I have made are highly personal and represent my ongoing interest not only in interface technology, but also in the philosophies surrounding new technology and cyberculture. For the sake of comparison approaches have been chosen that are in some cases diametrically opposed to each other.

For the purposes of this discussion, alternative interfaces can generally be described as computer input systems that allow the user or users to connect to a digital environment without the use of conventional keyboard-and-mouse interfaces. More specifically, these systems allow

Perspectives on Multimedia R. Burnett, Anna Brunstrom and Anders G. Nilsson
© 2004 John Wiley & Sons, Ltd ISBN: 0-470-86863-5

users to connect to and/or enter an artificial environment, and to control multimedia elements (text, images, video, sound, and data) when in that environment.

The most popularized examples of alternative interfaces include systems for the handicapped (i.e. eye-tracking systems) or systems using a 'touch-based' input (i.e. touch screen kiosks). While these systems certainly provide an 'alternative' to the existing mouse–keyboard dominance, for this discussion we will limit ourselves to more obviously 'experimental' prototypes.

Many of these experimental systems are of a unique design, specially created for the needs of one or more persons. Others use existing technologies to interface with a digital system in previously unthought-of ways. In almost all cases these systems are in the 'research and development' stage, are created for one personalized user, or have yet to reach mainstream popularity.

2.2 CYBERNETIC SYSTEMS

Cybernetic systems may be described simply as systems in which the user is physically connected to a computer and/or digital environment. While all of us who use the mouse–keyboard interface participate in a superficially 'cybernetic' relationship with our computers, true cybernetic systems go far deeper in their attempt to connect the human form with an electronic environment. In any attempt to connect humans more intimately with technology, an actual physical interface is required between man and machine. The most extreme, and therefore obvious, example of this sort of cybernetic system is the technology developed by Stelarc, the Australian performer, artist, and the world's first genuine cyborg.

In short, Stelarc has invented a system that allows a user (namely himself) to control intimately, and be controlled by, a number of electronic devices. The complexity of the human–machine interaction and the number of devices Stelarc controls and is controlled by is rather daunting, so for the purposes of simple comprehension we will talk about one device, the *Third Hand*.

Figure 2.1 Stelarc, *Third Hand*

With the *Third Hand* Stelarc uses electrodes attached to various portions of his torso and upper thighs to connect to a third hand (see Figure 2.1). In his own words:

> The artificial hand, attached to the right arm as an addition rather than as a prosthetic replacement, is capable of independent motion, being activated by the EMG signals of the abdominal and leg muscles. It has a pinch-release, grasp-release, 290-degree wrist rotation (clockwise and anti-clockwise) and a tactile feedback system for a rudimentary "sense of touch." Whilst the body activates its extra manipulator, the real left arm is remote-controlled / jerked into action by two muscle stimulators. Electrodes positioned on the flexor muscles and biceps curl the fingers inwards, bend the wrist and thrust the arm upwards.
>
> (Stelarc, 1999a)

If one looks closely at the picture of the Third Hand reproduced in Figure 2.1, one can see cables attached to electrodes on Stelarc's stomach and thighs. These cables are passing information transmitted through the stomach and thigh electrodes. The information is produced by the specific contractions of Stelarc's muscles (an amazing feat in itself), thus creating a feedback system between the physical actions of the human body and the programmed responses of a robotic device. Throw in the fact that the arm can send pulses to the attached body in order to trigger involuntary muscle response and you have in essence the first working example of an operational cybernetic system.[†]

The stated goals of Stelarc's systems are extremely wide ranging, and thus rather open to criticism for their apparent pretentiousness. In essence, Stelarc seeks to update the human body for the twenty-first century, to create a new evolutionary leap for humanity that will allow us to keep pace with the rapidly expanding powers we have given to our machines. In Stelarc's case, the question is not *whether* we need to enhance the human–machine interface to include more intimate connections, but *how* we will do it. To that end, Stelarc has been actively involved in the creation of devices such as the Third Hand, which go a long way to bridge the physical gap between man and machine. Whatever one may think about the portentous philosophical claims made by Stelarc in his writings,[‡] there is no doubt that what he is proposing is a viable interface system for at least one user: himself.

As far as an approach to interface technology is concerned, Stelarc's systems, while obviously functional, are at the current stage very limited in their potential user base. A device such as the Third Hand, with it's possibility for remote actuation of a body and it's actual physical connection to the human torso, obviously implies a certain approach toward interface design, and that approach is, in a word, 'invasive.' Certainly it seems apparent that not many humans are ready for the level of invasiveness needed to employ Stelarc's cybernetic systems, or even

---

[†] Right at the outset, many people have a difficult time accepting that such a system even exists and is being used. Invariably when I show this work, audience members are certain that this *must be* science fiction and cannot possibly be *real* given our current scientific reality. For those of you who need convincing and have Web access, connect to 'Stelarc Official Website', <http://www.stelarc.va.com.au/third/third.html>, and click on 'Third Hand as navigable vrml video.' This should provide some documentation of the system.

[‡] See the Stelarc web-site, <http://www.stelarc.va.com.au/>, for examples of his writing as well as photos of his various cybernetic devices.

'lighter' cybernetic systems such as Steve Mann's wearable computing (see Mann, 1999).

The invasiveness of Stelarc's systems, and by extension much work in cybernetic systems, is sustained by an overriding philosophical belief in the value of *intimately* connecting humans with their machines. In the case of Stelarc, the cybernetic link is so complete that one can in fact begin to speak of him as a cyborg, that is, a being who is human with technological *additions* to his body. Certainly to anyone who has witnessed Stelarc in performance, it appears as if the system is cybernetic, invasive, and even *encumbering* (the Third Hand weighs *ca* 20–30 kg). In its present state this sort of invasive system is certainly not for everyone. As stated above, most computer users are simply not ready to accept this degree of cybernetic invasion, nor are they physically adept enough to manage such a complex and physically demanding system. Stelarc insists, however, that we must prepare ourselves for the day when such a system may be essential for survival:

> It is time to question whether a bipedal, breathing body with binocular vision and a 1400 cc brain is an adequate biological form. It cannot cope with the quantity, complexity and quality of information it has accumulated; it is intimidated by the precision, speed and power of technology and it is biologically ill-equipped to cope with its new extraterrestrial environment.... The body's LACK OF MODULAR DESIGN and its overactive immunological system make it difficult to replace malfunctioning organs. It might be the height of technological folly to consider the body obsolete in form and function, yet it might be the height of human realizations. For it is only when the body becomes aware of its present position that it can map its post-evolutionary strategies. It is no longer a matter of perpetuating the human species by REPRODUCTION, but of enhancing male–female intercourse by human–machine interface. THE BODY IS OBSOLETE.
>
> (Stelarc, 1999b)

While there exists a substantial degree of theatricality in this and other examples of Stelarc's pronouncements, there is a rather 'grounded' aspect to his work as well: these are not mere dry speculations, but theories borne out in the practice of his design strategies. Stelarc is in a sense offering himself as the first experimental test subject of a cybernetic human–machine interface. This willingness to experiment on himself (rather than on others or on animals) undermines some of the knee-jerk reactions to his work, which often predictably claim some sort of 'superman' fascism lurking in Stelarc's theory and practice (as if he was forcing everyone to wear his devices). On the contrary, Stelarc insists that freedom to pursue these aims is an individual choice and should be allowed to flourish in many forms:

> In this age of information overloads, what is significant is no longer freedom of ideas but rather freedom of form—freedom to modify and mutate the body. The question is not whether society will allow people freedom of expression but whether the human species will allow individuals to construct alternate genetic coding.

> (Stelarc, 1998, p. 117)

Additionally, the implications of Stelarc's devices go beyond the rather limited performance-based applications that they have been currently restricted to. Stelarc himself has suggested that the Third Hand technology should be used to create a new generation of accelerated learning tools. In essence, one user would use the Third Hand or some variant of it to instantiate a second user into motion (though control should go both ways according to Stelarc), thus enabling the teaching of motor skills by remote activation:

> Technology now allows you to be physically moved by another mind. A computer interfaced MULTIPLE-MUSCLE SIMULATOR makes possible the complex programming of either in a local place or in a remote location. Part of your body would be moving; you've neither willed it to move, nor are you internally contracting your muscles to produce that movement. The issue would not be to automate a body's movement but rather the system would enable the displacement of a physical action from one body to another body in another place—for the on-line completion of a real-time task or the conditioning of a transmitted skill. There would be new interactive possibilities between bodies. A touch screen interface would allow programming by pressing the muscle sites on the computer model and/or by retrieving and pasting from a library of gestures.... THE REMOTELY ACTUATED BODY WOULD BE SPLIT—on the one side voltage directed to the muscles via simulator pads for involuntary movement—on the other side electrodes pick up internal signals, allowing the body to be interfaced to a Third Hand and other peripheral devices. THE BODY BECOMES A SITE FOR BOTH INPUT AND OUTPUT.

> (Stelarc, 2000, p. 121)

Stelarc's alternative interface design can be summed up as a highly responsive and interactive system tailored for one user, adaptable in the long term for a small number of users, invasive, encumbering, but capable of enormous control and feedback between attached humans and machines. The system has tremendous potential applications for the training of motor skills, for enhancement or improvement of human sensory input, and (although Stelarc doesn't intend this) by providing replacement limbs for people who have suffered traumatic dismemberment. However, while Stelarc's Third Hand system is extremely compelling and fascinating, it's invasiveness in it's current form makes it a rather specialized tool, one that is unlikely to discard the term 'alternative' in favour of 'mainstream' for quite some time.

2.3 'HANDS-FREE' TRACKING SYSTEMS

In rather direct opposition to the invasive and encumbering nature of Stelarc's devices, there exist several interface technologies that afford the user an intimate and personalized interface to digital systems without the need for any cybernetic additions to the body. In fact these systems go the opposite route from Stelarc by trying to hide evidence that there is any interface at all.

Although there exist many types of 'hands-free' tracking devices, for the sake of simplicity we will focus on one such system, the *Martin Lighting Director* (MLD), invented by Will Bauer. With the MLD one is immersed in a system where the displacement of a body in space triggers events in a digital environment; however with the MLD the interface that controls the human-machine interaction is not added to the body but is literally accessed 'in the air.'

The MLD uses an ingenious and clever application of an existing technology: *sonar*. In submarines sonar is used to calculate depth by sending high-frequency signals to the ocean floor and thereby establishing depth by calculating the time delay the signal takes to return to the source transmitter. With the MLD four sonar transmitters are placed in a large room (up to approx. 400 m^2). Each sonar transmitter sends pulses 15–60 times a second at a frequency just above human hearing (*ca* 20 kHz). The user wears a small ultrasonic wand that weighs less than 250 g. This device is so small and unobtrusive that it can be strapped onto the user's belt or arm. The user has very little sense that he or she has any addition to his or her body and therefore the system is, in essence, 'hands-free' and the user is unencumbered.

As the user walks through the room bounded by the four sonar transmitters, the ultrasonic wand receives pulses from the transmitters. The wand then calculates where the user is standing in relation to centre of the room. The coordinates are given in three dimensions; left to right (x), front to back (y), and floor to ceiling (z). The centre of the room on the floor measures 0,0,0. Movement to the left of the centre produces negative coordinates on the x-axis, movements to the right produce positive coordinates. Movements forward from the centre produce negative coordinates, in centimetres, on the y-axis, movements backward produce positive coordinates. Movements from the floor to the ceiling produce positive coordinates on the z-axis.

As the user moves through the room, the coordinates are updated 15–60 times a second (the larger the room, the lower the sampling rate). The coordinates are then sent via radio to a receiver connected to a PC. The PC receives the coordinates, and sends those coordinates to an interface software. The system allows a user simultaneously to control protocols such as DMX512 (for lighting), serial and UDP data output (for multimedia and VR) and MIDI (for music). The software can be configured so that audio, video, animation, and lighting effects can be 'located' at specific coordinates in the room. As Figure 2.2 shows, with a MLD wand the user can access specific images at points in a room and move the images around simply by walking. As the user moves, sound can also be made to fade in or out in real time or perhaps to change in pitch. Any parameters that are controllable by MIDI (the language used by computers to describe musical events) can be altered in real time by the user. Similarly, images, lighting effects, and even QuickTime movies can be placed in different parts of the room, and accessed and run by the movements of the user. The interface can also be configured to give the illusion that as the user walks, he or she walks through an audio-visual environment. This environment is usually projected onto a large screen or inside a virtual reality helmet. The user therefore experiences what it would be like to move through an imaginary environment.

It goes without saying that this system affords the user considerable freedom of movement, and requires no extraordinary feats of muscle control, other than walking. In this way the MLD and other 'hands-free' systems, represent the opposite pole of design thinking from Stelarc:

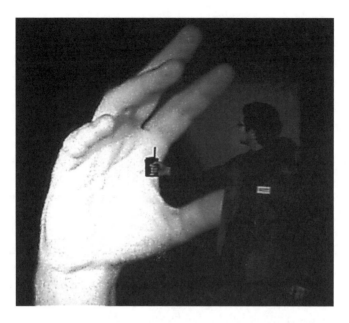

Figure 2.2 The user moves the ultrasonic MLD wand and is followed by a
huge projected grabbing hand. From *Idle hands,* by Rafael Lozano-
Hemmer and Steve Gibson

they are in no way invasive, and can be used by just about anyone
(rather than one specialized user). These systems also proceed from
the belief that current mainstream interfaces are limiting, potentially
unhealthy (we sit and flap our arms all day long at a keyboard), and afford
very limited control of an environment. The MLD is complex enough
to be customizable for a huge number of applications, simple enough
to be preprogrammed for use by children or adults with no computer
background, and responsive enough to allow anyone to experience the
illusion of moving through a system.

The MLD has obvious and wide-ranging applications in areas such
as gaming, lighting for theatre and rock shows, large-scale 3-D architectural design, installation art, and virtual reality. This, and other hands-
free systems, represents a truly alternative approach to interface technology: unique, user adaptable, fully immersive, and yet not requiring
you to become a Borg. The crowning achievement of the MLD is that it
can accommodate several levels of user expertise, from the completely

uninitiated to the expert. In this way, hands-free systems represent the next logical step for interface technologies—into the realm of the tactile. Their flexibility makes them the most likely alternative interfaces to make their way into the mainstream.

2.4 MIDI INSTRUMENTS AS VISUAL TRIGGERS

Continuing in the vein of existing technologies used for purposes other than originally intended, we now come to MIDI instruments. Just as the *Martin Lighting Director* uses sonar technology in a previously unexplored manner, so MIDI technology can be used to control an environment it was not intended to be used with.

MIDI (Musical Instrument Digital Interface) is both the interface and the language used by computers and MIDI instruments (keyboards, synthesizers, drum machines, samplers, sequencers) to transmit and convey musical information back and forth. Information on what note was hit, how hard it was hit, and all the knobs, wheels or pedals that were moved on a keyboard can be sent to a computer as textual information. That information can be stored by the computer in a type of software called a *MIDI Sequencer*, and can then be played back to the synthesizer at a later time. This procedure can be repeated for all the MIDI instruments in a studio, thus allowing a single user to build up a layered musical composition. In addition, the MIDI data can be edited non-destructively (i.e. without damaging the source material) on the computer, thus allowing the musician to correct mistakes, add embellishments, create volume swells, etc. If this seems too hard to understand to the non-musician, consider a different metaphor: a MIDI keyboard is like a player piano, and the MIDI sequencer is like the punch-card that goes into that player piano. As a sort of 'digital' punch card, the MIDI sequencer faithfully replays what was programmed into it originally.

The beauty of MIDI is that no actual *audio* information is passed between the synthesiser and the computer[†] (similarly the punch card on a player piano doesn't contain sound, only information on how to play

---

[†] Digitized *audio* is a huge consumer of computer memory. Uncompressed, one minute of stereo CD quality audio would take up 10 megabytes of storage space.

the sound). In a MIDI set-up, all that is passed back and forth between the synthesizer and the computer is textual information describing *how* to play the sound contained *on* the synthesizer. For example, the hardness with which you strike a note on a keyboard is described as *MIDI Velocity* and is measured between 1 (very light touch) and 127 (very hard touch). Usually this parameter is used to make a note *sound* hard or soft (although it can be mapped to other musical parameters as well). When a *velocity* of 127 is recorded to a MIDI sequencer and then played back to the original synthesizer it will probably sound very loud. All that is transmitted between the sequencer and the synthesizer is a couple of numbers and not complex digital audio consuming megabytes of hard disk space. Therefore, even with multiple MIDI instruments playing several tracks, MIDI file sizes tend to be very small (usually the amount textual information transmitted in an orchestral-sized MIDI file is much smaller than the Microsoft Word document used in the creation of this article). I have composed large-scale MIDI compositions of up to 20 minutes in length, and the file size was as low as 196 K!

Therefore, the obvious question arises: as MIDI is simply a textual language, could it not be adapted for various non-musical uses? In fact MIDI is already being used by various individuals for the control of environments that may include not only music, but also other media in synchronization.

Several software producers have been involved in the creation of MIDI Xtras; in essence, these are little MIDI software routines that allow a user to control multimedia elements through a popular piece of software called macromedia *Director*. In general, Director is used to produce CD-ROMs or web sites in which the user interactively navigates with a conventional mouse–keyboard interface. There is no reason, however, why some other interface device, including the Martin Lighting Director, a virtual reality helmet, or input from a MIDI instrument or instruments could not facilitate user interaction in Director. The simplicity of MIDI instruments, their ubiquity, and their ease of use (at least for musicians) makes them ideal interface devices for navigating a multimedia project, though for rather strictly limited purposes. In the alternative interface world, MIDI instruments can be termed as expert interactive systems that require specialized knowledge, but do not require especially encumbering or personalized devices.

2.4.1 Visual music

Visual Music is the theme of a series of pieces I have created which seek to find a database of programming relationships between musical data, as represented by MIDI information, and visual links, as represented by controllable animated sprites. The goal of my performance pieces *Cut to the Chase* and telebody has been to define a series of programming behaviours that allow the musical performer to extend his or her hand into the control of visual objects. In collaboration with Jonathan Griffiths, we have developed a database of possible relationships between audio and visual media. Instead of relying on older 'synthaesthetic' models (e.g. colour-based) of visual music, we have instead tried to find obvious musical–visual corollaries that seemed logical: stereo audio pan = screen position, keyboard velocity = image brightness. This has the advantage of creating seamless and obvious musical–visual relationships for the viewer, creating a true rhythmic experience of visual music.

2.4.2 Visual music part I—cut to the chase

I have been involved—with Bert Deivert and Jonathan Griffiths—in research to standardize the use of MIDI as an input device for controlling visual environments. This work has been completed using the programming language called *Lingo*, which is used to create user interaction in Macromedia *Director*. Employing one of the available MIDI Xtras for Director, we have been creating Lingo programming scripts, which enable a user to trigger and control a visual environment using standard musical data that all musicians employ when playing musical instruments.

The goal was to make the musical interface (any MIDI instrument) seamlessly control a visual environment to the point where the musician would be able simply to 'play' the images as if he or she were 'playing' musical notes. In short the MIDI-to-visual interface should be 'invisible' enough that the musician can comfortably use techniques that he or she commonly uses when performing music, thorough enough for subtle variations in musical playing (i.e. hitting a note hard or soft) to produce clear visual corollaries (a lighter or darker image is triggered), and obvious enough for spectators to understand what is happening in the process.

The initial result of this research into the use of MIDI Instruments as an interface to control visual environments was explored in our audio-visual performance piece *Cut to the Chase*, which toured extensively in Northern Europe in 1997 and 1998. The piece was successful as an example of a MIDI–visual interface in many regards, though in some cases it was apparent to us that improvements needed to be made before moving further with this system as a viable interface technology. It is clear to us that the interface is still in the research and development stage, and that at present it is primarily relevant to a small (but we hope significant) group of expert users of interactive systems. Its primary applications are currently in live audio-visual performance, and real-time video editing.

The developers of the MIDI-Visual interface are philosophically committed to the idea of expert-user systems. While there is obviously a great need for alternative interface systems that can accommodate beginning users (e.g. the Martin Lighting Director), there also is a need for expert systems that allow advanced users to control an environment precisely, without having to resort to the encumbering interfaces needed in many cybernetic systems. Expert user-interaction systems can take many forms, and we are certainly not suggesting that the MIDI-Visual interface as employed in *Cut to the Chase* and *telebody* is the only possible approach, or even the best in this area. We simply offer it as an example of an alternative interface system as seen from the inside, through the eyes of the developers.

Cut to the Chase is a performance piece that utilizes a distinctly unique approach to multimedia performance. In *Cut to the Chase* video and animation are *performed* live. As described in the previous section, the MIDI-Visual interface is controlled by special programming that allows the user to play on a MIDI guitar in order to trigger and control images. We had several goals in the creation of this interface design, not limited to the creation of an expert user-interaction system for musicians. The various aims of the developers are outlined in the following paragraphs.

In both performance and film, video and animation are generally *fixed* media. In other words, when experiencing these media, the user tends to view a pre-recorded, previously edited and mastered tape. At a film screening, in a multimedia performance, and even on a CD-ROM, film and video 'performance' is usually limited to pressing the 'play' button. In contrast, in the 'old-fashioned' audio-visual media of theatre and opera, the visual materials rely on a live performance element, and to a certain extent, improvisation. What if video, film, and animation could be 'performed' live, 'scored' and/or 'improvised' according to the wishes of the film maker or video performer(s)? In *Cut to the Chase* this is what we have sought to achieve.

By implication, if the capability exists to 'perform' video live, then video can also be edited and rendered in real time, but not just edited in the conventional sense of matching cuts, but also as in adding effects and synchronizing audio-visual elements. Using this technique to edit video in real time, users could potentially edit multiple takes of video scenes, with no need to wait for long rendering cues. This possibility is implied by the technique in *Cut to the Chase*.

In *Cut to the Chase*, a musical performer is able to improvise images and video in real time. The performer can simply play notes on his MIDI instrument and Director will respond by calling up specific image and/or video files—with or without backgrounds. Additional manipulation of the images and videos is available by varying the playing technique on the performer's MIDI instrument. For example, striking notes harder or softer will change specific parameters such as playback speed, playback direction, image brightness, or screen position of the selected image or video. Logical connections are made between performance technique and the resulting visual effect. For example, when a note is struck softly the related image will be dark, but when the note is struck harder the image will increase in brightness. Additional commands mapped to specific musical parameters (pitch bend, note order, velocity) allow for basic control over most of the video and image features enabled by *Director's Lingo* scripting language.

Due to limitations of time and to the newness of the technology, in *Cut to the Chase* we restricted the number of video performers to one, and we allowed this performer to trigger images only, not sound. The video performer 'accompanies' a musical performer by 'playing' image sequences in response to the musical materials. The fascination of the process is that the images are spontaneously accessed and edited, allowing for visual improvisation, imitation of musical rhythms, and performer errors: in essence all of the elements necessary for true real-time performance.

The foremost objective of *Cut to the Chase* was to use compositional/improvisational techniques familiar to musicians in order to control a visual environment. This has the advantage of using something familiar to many people (musical technique) to do something new (trigger and alter visual images). The second objective was to allow real-time video performance along with the musicians. This real-time video performance allows for variations in tempo and dynamics not normally possible with video playback. In addition, exact timing correspondences between the audio and visual realms can be achieved in real time, rather than by pre-editing. The third objective was in fact an afterthought and is a by-product of our research. With the interface technology and the standardized music-imaging programming in place, video editors could potentially use a small MIDI keyboard attached to their computers to allow them to edit and synch audio and visual materials. No long rendering times would be necessary, and these users would be able to improvise multiple edits of an image sequence and choose the best one.

2.4.3 Visual music part II—telebody

After performing *Cut to the Chase* for two years and therefore becoming intimately familiar with all the problems and possibilities associated with the MIDI-Visual interface, we decided to embark on a new project called *telebody*. In *telebody* each performer is allowed control over MIDI-controlled sounds as well as MIDI-controlled images. This idea came from our realization that for this interface to be successful as a performance tool, viewers must *clearly* understand what is happening in the process: they must recognize that there is an obvious connection between a musical performance and the corresponding visual result. In *telebody* the performers control a very small body of musical materials (one vocal melody for example) which has obvious links to the visual materials (the vocal melody may trigger the movements of a female body). In essence the idea is to establish a norm for working in a multimedia performance environment, to create a recognizable medium for the viewer (in the same way that film and rock video have standardized forms that you and I recognise, but would be meaningless to a Nineteenth-century viewer). The interface used should be capable of enhancing viewer understanding of the medium, while affording the performer new and unique ways of creating and controlling an audio-visual environment. In this sort of scenario, precise control over the synchronization of visual and musical elements will be available in the hands (literally) of the performers.

The original aim of Visual Music as explored in *Cut to the Chase* was simply to get images to trigger in response to musical information: a note plays and an image appears. As we explored the technique further, it became apparent that control could go a lot further than this simple, rather banal technique. Interaction between the musical and the visual could be defined in a series of changing or fixed relationships between musical values and visual results. Even real-time form, in the musical sense, could be programmed to be understood in the visual realm, with new groups of images called up in response to musical cues. As we developed telebody we began to realize that the number of corollaries between musical information and visual result was much broader than we had originally thought.

Using *Troika Tronix's* excellent MIDI Xtra for Director called XMIDI (Troika Tronix, 2003), Jonathan Griffiths and I set out to explore *total* control of a visual environment by MIDI instruments. Our goals were technical, philosophical and aesthetic. In short, our idea was to make an application that reflected control of the image by another's musical action. Human bodies, manipulated and/or natural, were chosen as the visual material with which the musician could interact.

In our programming, several different series of animated still images were hard-wired to musical leitmotifs or sound sources. A clear female voice was matched to a spinning animation of a female body, and as the sound of the voice was manipulated, the image was manipulated in turn. Simple relations for real-time manipulation were established for each hard-wired pairing: using a particularly intense analog sound to trigger a dramatically manipulated male body, we wired pitch bend to image skew, allowing the performer radically to stretch the sound and image in synch in real-time. As these pairings unfolded in time, a narrative sense emerged, based on marrying of a rather hard, scientific database rela-tionship between sound and image, and the raw physical act that these pairings resulted in. The sense that the bodies were moving without personal agency, but rather triggered and manipulated by external musi-cal control, created a strange puppet dance that unfailingly retained its

Figure 2.3 Steve Gibson and Bert Deivert performing telebody at Open Space Gallery, Victoria, Canada. Photo by Rachel Keirs

master–servant synchronization. As Figure 2.3 shows, the performer on the left is controlling one body with the keyboard and the performer on the right is controlling a second body with the guitar. The image in the centre is controlled by the MIDI sequencer playing an arpeggied melody.

As our programming developed, it became clear that certain music–image control relationships had logical corollaries that could be maintained through the course of the piece. So just as certain animated still images were hard-wired to musical leitmotifs, certain musical control functions were mapped to certain image control functions. While we made no pretensions to scientific exactitude, these pairings seemed logical enough and for the viewer they began to make sense with repeated viewing. Table 2.1 lists the established control pairings that hold true for the most part during *telebody*:

This combination of musical-to-image relationships allows the musician to predict an outcome of his or her musical actions in the visual

Table 2.1   Control pairings in *telebody*

MIDI event	Visual result	Lingo code
1. Note On	Animation or still image triggered.	AnimMod or StillMod
2. Pan	Image horizontal screen position.	LocH
3. Note velocity	Image brightness	Blend
4. Pitch bend	Image Skew	Skew
5. Modulation wheel	Image Rotation	Rotation
6. Controller 99	Image cast to be used	ModPos

realm. Figure 2.4 for instance shows a body being pulled apart by the notes played by the guitar player. The visual figure splits into a cross-like object depending on note velocity and is precisely matched to a sample triggered by the guitar of a boy's choir singing '*Agnus Dei*'. Through this equating of image to sound, the viewers begin to establish that musical actions have exact visual corollaries, giving them a strange sense of the synthaesthetic power of the musical performer.

2.4.4   Visual music part III—aesthetics and ideology

Our overriding goal in *telebody* was to explore how this technique could be used to create a unified and satisfying artistic experience, one where form and content would be almost dictatorially matched. We were interested in exploding meaning from the crash of musical performance and body manipulation. There was no specific intention to make an exact ideological statement with the piece. The ideology emerged as if naturally from the perfection and reliability of the technology itself. The bodies unfold perfectly in synch as if scientifically controlled; they assume new forms alternatively lovely, grotesque, strange, overactive, impossible, and all the while they remain without any personal agency, continually manipulated by their external controllers. They have lost the ability to act on their own personal sense of agency; they are at the mercy of the technology.

In this manipulation there lies a theme that has dominated my work from the time of *SPASM* (Kroker and Gibson, 1993): the relation of the

Figure 2.4 Bert Deivert performing telebody at Open Space. Photo by Rachel
         Keirs

body to technology and its increasing loss of defining separation from
that technology. I have always approached this theme with a certain
amount of detachment and a lack of hysteria. For me, the fact of technological enhancement and interface simply *exists*. Naturally there are
social, political and cultural implications, but my interest has always
been piqued by the thought of what new forms could be conceived. As
an artist I was interested in exploring those forms for their own sake: what

would a super-male voice sound like, and how about a hyper-female? How does gender dissolve in the sampling world? Finally, what do digitally manipulated bodies look like? In *telebody*, the brutal dissolving of the human form became the ideological component of the piece, and in the end it is both celebrated and bemoaned, often simultaneously. The ideology of cybernetic control of the body is coolly calculated in *telebody*, precisely programmed, mapped out and performed. The performers become the biogeneticists, with no ethical limits placed on their behaviour, no thoughts about what they are doing, just the sense of power that they are doing it.

In the final analysis the question of biotechnology seems inevitable, and certainly the mode of performance in *telebody* exploits the notion of performers as mad scientists (though in quite a different sense than explored in Perry Hoberman's piece *Let's Make a Monster;* Hoberman, 2003). In *telebody* the performers don their white suits and manipulate the bodies into submission; however, there is more here at work than simple manipulation. Underneath there is an emotional core that reaches a fever pitch at points. The pummelling loudness of the images and the music in *telebody* lends a sort of romantic doom to the piece, and in essence the core of the work is melancholic, but only to a point. The end is meant to be a final classical orgasm of release: male completion at the end. The mad male scientist fantasising about new forms of control. In this way *telebody* does have a certain critical air to it, but it is ambivalent: it is curiously attracted to the new forms, it relishes the control it has, it is emotionally wrought, but at the same time a nagging feeling of Orwellian paranoia hints in the back of its mind. In this way *telebody* captures both the formal perfection engendered by its technology, and the pleasure and pain such control could exert on the human form.

2.4.5 Visual music coda—a note on the body

The body of the performer is of paramount importance in *telebody*. While it is only tracked in order for the robot lights to follow it, the body of the performer is physically involved in the manipulation of the virtual body. Its hands pummel the bodies into submission, its dancing is a puppet celebration of the moving bodies, occasionally even its voice causes the disembodied heads to sing. Using the real body to manipulate another virtual body engenders a satisfying formal unity: simple correlation between form and content, full circle.[†]

2.5 CONCLUSIONS

This chapter has served as an introduction to the world of alternative interfaces. As stated at the outset, most alternative systems are experimental in design and are thus difficult to discuss from any *standardized* point of view. These systems also attempt in one way or another to create a more immersive environment for the user, thus creating a more complex interaction model than any used in mainstream interfaces. Proceeding from different philosophical starting points as these systems do, they nevertheless manage to establish cogent and workable theoretical and practical models for widening our sensory interaction with digital environments. Indeed, perhaps one of the major problems with mainstream technological systems for most ordinary users is that they are *too* standardized, and thus at worst useless for very specific tasks, and at best completely impersonal as communication devices. The systems outlined in this paper are seeking various ways to create user-interaction systems that are more responsive to the tactile systems of the human body, to bring the human into greater contact with, and thus control, a digital environment. They acknowledge, in their design aim, McLuhan's fateful predictions that those of us who do not understand technology—and by extension those who do not master and control technology—will be enslaved by it.

Whether or not these systems will see the light of day in a wider milieu is difficult to predict. Each of the above approaches represents a different target group, from the extreme, niched user base of Stelarc's systems, to the expert user base of the MIDI-to-visual interface, on to the wide-ranging user base of the Martin system. Each of these systems represents a different philosophical and technical approach to the problems of working with more intimate physical interface technologies.

[†] Rafael-Lozano and I explored a similar technique in *Idle Hands,* where a users hands controlled virtual hands, which were in turn manipulated by another user's hands.

Each of these systems has its strengths and weaknesses depending on the application of the end-user. In the final analysis, one can only hope that this sort of creative research in the area of machine–machine relationships will offer a wider range of possibilities than that allowed by the present mainstream interface.

2.6 REFERENCES

Acoustic Positioning Research (2003) Acoustic Positioning Research web-site, <http://www.positioning-research.com>

Hoberman, P. (2003) Let's Make a Monster!, <http://www.perry-hoberman.com/monster.html>

Gibson, S. (2003) telebody web-site, <http://telebody.ws/>

Kroker, A. and S. Gibson (1993) *SPASM: Virtual Reality, Android Music, Electric Flesh,* St. Martin's Press, New York.

Lovejoy, M. (1997) *Postmodern Currents: Art and Artists in the Age of Electronic Media,* Prentice Hall, New York.

Mann, S. (1999) 'Cyborg Seeks Community', *Technology Review,* 102, (3) 36–42, or <http://www.wearcam.org/mann.html>

Mann, S. (2003) Prof. Steve Mann web-site, <http://www.wearcam. org/mann.html>

Martin (2003) Martin Lighting Director page, <http://www.martin. dk/product/ product.asp?product=lightingdirector>

Stelarc (1998) 'From Psycho-Body to Cyber-Systems: Images as Post-Human Entities', in Dixon, J. and E. Cassidy (Eds). (1998) *Virtual Futures: Cybererotics, Technology and Posthuman Pragmatism,* Routledge, London.

Stelarc (1999a) Stelarc Official Website, <http://www.stelarc.va.com. au/third/third.html>

Stelarc (1999b) Obsolete Body, <http://www.stelarc.va.com.au/obsolete/ obsolete.html>

Stelarc (2000) 'From Psycho-body to Cyber-systems: Images as Post-human Entities', in D. Bell and B. M. Kennedy (Eds) (2000) *The Cybercultures Reader,* Routledge, London.

Stelarc (2003) Stelarc, <http://www.stelarc.va.com.au/>

Troika Tronix (2003) XMIDI, <http://www.troikatronix.com/xmidi.html>

3

Transparency, Standardization and Servitude: the Paradoxes of Friendly Software

Andreas Kitzmann
Media and Communication, Karlstad University

3.1 INTRODUCTION

The world of multimedia development is a fickle one. Yesterday's visionaries can quickly become today's lunatics and the decisions regarding 'what's hot/what's not' appear to be about as rational as a coin toss. Accordingly, bad ideas can be as successful as very good ones, perhaps even more so judging by some of the products on the market today or, conversely, by those that have regrettably faded away.

Looking back at the heady days that preceded the spectacular rupture of the digital bubble, I am brought back to a presentation that I

Perspectives on Multimedia R. Burnett, Anna Brunstrom and Anders G. Nilsson
© 2004 John Wiley & Sons, Ltd ISBN: 0-470-86863-5

had attended at MIT some years ago which, in the context of today's economic and technical climate, bears a number of useful insights. The speaker was Bob Stein, one of multimedia's early pioneers, whose now derailed Voyager company was one of the first to attempt to bring multimedia publishing to the consumer market. At the time of his presentation, Stein had already left Voyager and was eagerly pushing his next Big Idea that went by the rather unlikely name of *Night Kitchen*.

The stated goal of the project was fairly straightforward: to produce a multimedia authoring program designed to allow users with modest computer skills to produce professional looking and structurally complex works of multimedia. These multimedia documents could take on a variety of forms—from personal web sites to interactive novels to fully blown multimedia presentations on CD-ROM with text, sound and video files.

Stein's aim with Night Kitchen is essentially to democratize the field of multimedia authoring by reducing the amount of specialized knowledge necessary to produce a multimedia work competently. Multimedia production would, according to such an aim, become a normative medium for expression in the sense that it would gain greater accessibility among the wider public and thus be more apt to enter into the realms of everyday life.

One of the aims of this article is to place the ambitions of Bob Stein within a larger discursive context, focusing in part on the major assumptions and paradigms that lie behind the development and applications of multimedia. In this respect, my interests lie not in evaluating the particular technical details of Night Kitchen in terms of its merits or faults, but rather in using the program and its developer, Bob Stein, as a representative example with which to explore a number of cultural, ideological and theoretical vectors. Specifically, I wish to address how Stein's concerns for democracy and expressive freedom are indicative of the discourses of emancipation and enlightenment which continue to resonate within the cultures of the computer.[†] Such discourses can be seen for example, in the commitment to hypertext as a means with which to overcome

---

[†] The term 'democracy' should be understood in its most general sense. In the case of its use by the computer industry 'democracy' is often defined in terms of interactivity, easy access to technology, user friendliness and trouble-free communications.

the so-called limits and tyranny of the printed word by virtue of being a technical medium which better reflects the nature of our consciousness and expressive needs. Equally pervasive is the juxtaposition of the spiritual (or metaphysical) with the technological which characterizes cyberspace as a vehicle for transcending the grimy complexities of the physical world and entering into a realm of pure mind and spirit that knows no limits or restraints.

An additional and related concern involves Stein's ambitions regarding the normalization of multimedia production, and the manner in which his 'noble project' is potentially undermined by a paradox which often haunts those who wish to use the engines of mass consumption (or commercial culture) as a means for personal enlightenment. In this case, the inevitable result of making software easier to use is the increased reliance on standardization and pre-formatted options—a process which arguably limits the range of creative choice and expression. The democratic project of a program such as Night Kitchen is thus compromised before it even makes its way onto the user's desktop. What is intriguing about this dilemma is not so much the technical details but rather the manner in which the situation as a whole resurrects the much discussed (and yet to be resolved) argument regarding the effects of standardization and mass culture on human creativity. Also this argument crosses over into our total relationship with technology itself—a relationship that frequently skirts the multiple borders of suspicion, faith and outright obsession.

In addition to these topics and concerns, this chapter also reflects on the specific challenges involved in addressing computing technology from a critical and theoretical perspective. One of the major concerns here involves the question of how much technical expertise is required by humanities scholars in order to engage effectively with computing technology from critical and cultural perspectives. With reference to similar debates in the area of science studies, a collaborative approach is urged, which will bring the arenas of technique, theory and history into closer proximity. Such a collaboration, I believe, is important in terms of understanding technology through the perspectives of cultural and historical developments and, in this case, as a means to address and problematize the logic and assumptions that drive particular forms of software development and practice.

3.2 DIGITAL PROMISES

Night Kitchen is not an isolated example of goodwill on the part of a socially conscious computer programmer but rather indicative of a running theme (or discourse) among computer visionaries and developers (see Dery, 1996). One needs only to think of Theodor Nelson, the self-styled pioneer of hypertext and his mantra that 'computers are about human freedom' to get a sense of this major discursive strain or attitude. In a similar vein, the legendary and long defunct cyber-culture magazine, *Mondo 2000*, had, from its inception, combined advanced telecommunications technology and software with drugs and a kind of mystical eroticism that draws from a range of religious and philosophical traditions. In both cases, the intended result is a kind of 'cybertopia' where all desires can be immediately fulfilled (Taylor, 1999).

Equally celebratory are the theorists and practitioners of hypertext and hypermedia who claim that the non-linear, multitasking environments of the computer allow for entirely new feats of human imagination and expression—a claim that comes uncomfortably close to a form of technological determinism in the sense that it is based on the assumption that the technical nature of the medium is such that it can basically guarantee freedom. This capability is said to be due mainly to the fundamentally interactive nature of hypertext environments, which, according to Landow, allows readers to 'choose their own paths through bodies of information' and as a result 'find themselves with more power than might readers of similar materials in print'(Landow, 1999). Other critics, such as Sadie Plant go so far as to characterize the computer, especially the computer network, as a long awaited paradigm for realizing a world that is not bound by the constraints of gender and race. According to Plant, cyberspace is a 'naturally' accommodating environment for women by virtue of the fact that like women it is predicated on multitasking, fluidity, identity shifting and a decidedly non-linear perspective towards time and space (Plant, 1997). Thus, by implication, cyberspace becomes a potential vehicle for empowerment and self-expression. In a similar fashion, the various 'communities' within cyberspace, such as Usenets, MOOs and E-bay users have been depicted as important social formations that often manage to transcend the limits imposed by the physical world, such as those of gender, race and appearance (see Turkle, 1995 and Cohen, 1999).

What unites these claims and arguments is the assumption that information technology, especially the Internet and networked information spaces in general, somehow has a kind of built-in democracy. In other words, the form guarantees the function. However, as Foucault has reminded us, there are no machines of freedom. Liberty as always, is based on practice rather than form. As such utopianism ignores the very real and material conditions of cyberspace, which range from the infrastructure of the actual wires and machines to those of access and education.

Within the commercial world, the commitment to the transformative nature of computing technology is naturally more hyperbolic and insistent on the claim that computers can bring nothing but positive change. The Apple Computer Corporation, for instance, has long equated its particular approach to computer technology with the promises of empowerment, creativity and expressive freedom.[†] As such, being 'Apple literate' will allow users to reach higher and more fulfilling planes of expression, as indicated by the current slogan 'think different.' A visit to the company's home page reveals, for instance, that the iMac model released during the time of Stein's presentation is a veritable 'wish machine' which has the ability to make almost anything possible. The only limits, it seems, are those of the human imagination.

3.3 FIND WHERE EVERYTHING IS

> You'll find yourself discovering things you never knew existed. Shopping in places you've never been. Visiting the world's great museums and galleries. Researching projects for school or business. Reuniting with long-lost friends. Exploring the universe. The secret? Sherlock, the most powerful search technology on the Internet, and a key feature of the Mac OS (the system software your iMac comes loaded with).
>
> (Apple, 1999)

The publishing company Eastgate Systems, the developer of so-called 'serious hypertext' has a similar, but more restrained message—a message which captures some of the discursive spirit of hypertext

---

[†] The famous '1984' television commercial is a case in point. The current campaign of 'think different' continues this narrative of individuality and anachronistic creativity via technology.

scholarship.[†] Like the various Apple products, Eastgate provides tools that promise not only to unleash creative potential but also to take part in the revolutionary transition from an age of print to one of wholly digital media. Indeed, early proponents of hypertext publishing made much of this anticipated transition, claiming, as did Jay David Bolter in *Writing Space*, that 'the computer frees the writer from the now tired artifice of linear writing' in a manner that is liberating for both authors and readers (Bolter, 1991). One of the marketing campaigns deployed on the Eastgate home page plays on the common complaint that you can't read digital books in the bathtub while at the same time, asserting the message that digital writing is the way of the future and that any resistance is not only futile but reactionary. The main image of the site depicts two young children immersed in a bubble bath with one of them, a boy, looking directly into the camera with an expression of petulant boredom. Among the messages that float over the image is the equally petulant remark that 'we adore books, but fine writing is more than just paper and ink' and continues with a reminder that Eastgate has come up with a canon of its own.

> Since 1982 Eastgate Systems has been publishing some of the finest writing available anywhere—original hypertexts on floppy disks and CD-ROMS. So . . . GET OUT OF THE TUB!
>
> (Eastgate Systems, 1999)

What is notable about the examples of Apple and Eastgate is the implied desire by such companies to provide tools which, in one way or another, transcend the limits of human ability and thus provide tangible means with which literally to realize flights of fantasy. Such aims continue to provide one of the major narratives for the computer industry, thereby reinforcing the notion that computers are fundamentally about human freedom and that almost *any* engagement with such technology is likely to lead to some form of creative, material or even spiritual enhancement.

[†] Eastgate Systems is closely associated with the academic development of hypertext and hyperfiction. Many of the leading critics and authors of hypertext, such as George Landow, Michael Joyce and Jay Bolter, have worked closely with the company. The normative evaluation of hypertext is that it offers a challenging and productive alternative to the culture of print technology and is thus generally represented as providing users with an increased range of creative potential. For a general introduction to the hypertext and its theoretical/creative milieu see Landow (1994).

It is within such a discursive milieu that I would like to consider the aims of Bob Stein and his Night Kitchen project. Stein's goal of designing an authoring system that allows everyday users to create sophisticated multimedia documents can be read both in terms of confirmation and critique. On the one hand, Night Kitchen confirms the industry's message that computers are all about freedom—the freedom to create, to communicate, to express and to exchange. Such a message was made abundantly clear during the course of Stein's presentation. Yet on the other hand, Night Kitchen is also a critique of the perceived lack of usable multimedia software, at least as far as non-specialists are concerned. The program thus responds to the, as yet unrealized, promise of 'multimedia for the masses.'

Stein's sentiments here related to his earlier work with the Voyager company where he helped to develop the then nascent area of multimedia books, with an emphasis on publishing works that were perceived as having more cultural and historical than economic value. Voyager's expanded books, as they were called, were intended to be more than just passive consumer objects but rather vehicles for engendering meaningful dialogue, reflection and productive shifts in consciousness. Night Kitchen can be associated with such an agenda in the sense that it functions as a critical response to a perceived imbalance between multimedia production and consumption—an imbalance that potentially limits the scope and 'truth' of the 'digital revolution.' In other words, the ability to produce sophisticated and high quality multimedia works is still restricted in terms of cost and expertise, and is thus primarily the domain of professionals or at least advanced users. Higher-level multimedia authoring programs such as Director, are both more difficult to master and expensive, at least for the consumer market. In contrast, lower level programs such as Power Point are more accessible but generally do not allow the user much expressive freedom.

The main obstacle in such user-friendly programs is their apparent reliance on a higher degree of standardization with respect to design options, templates and tools—a standardization that is often necessary if the programs are to be 'friendly' with respect to cost and the time required to gain proficiency.[†] Night Kitchen is depicted as a direct response to

---

[†] Such an equation between ease of use and standardization is not entirely seamless. Indeed, one can make the case that many software packages aimed at the wider consumer market are increasingly

such a scenario. By utilizing the familiar 'drag and drop' mechanism for most of its functions, the program requires only a basic understanding of multimedia programming, layout and design while, at the same time, being said to provide enough flexibility to allow users to pursue their own creative urges to a fair degree of complexity. According to the company web site, Night Kitchen 'bridges the gap' that often separates users from developers, allowing even the casual computer user to create a variety of sophisticated and individualized multimedia documents. It is in this respect that Night Kitchen hopes to level the playing field of multimedia production.

During Stein's presentation at MIT, a question was raised regarding the paradox that results from the desire to create a software interface that is both accessible to the common user and that delivers a high degree of flexibility and individual control. It was noted that in the course of developing an authoring program that does not require highly specialized knowledge, it was usually necessary to standardize the options available to users in a manner that basically restricts how deep one can go into the program. One result is a kind of software signature in the sense that the user will find it difficult to create an aesthetic that is not overly based on the design parameters of the software package. In other words most projects created with Night Kitchen will look as if they have been created by Night Kitchen—a result which potentially undermines Stein's 'democratic project' in the sense that the standardization of creative output wins out over individual styles and formats. Such standardization is arguably less likely in higher-level programs because they provide both a range of more powerful tools and a greater degree of freedom with respect to programming and design options. The expense of such freedom, however, is not only a more expensive software package but, more importantly, the need for much higher degrees of expertise and programming knowledge. Bob Stein acknowledged this paradox and challenged those in attendance at the demonstration to come up with a better solution. The audience, as I recall, fell rather silent at that moment.

able to accommodate both new and expert users. The popular web authoring program Dreamweaver is one such example. The basic elements of the program are simple enough to learn quickly and thus allow most users to be able to design a web page shortly after removing the shrink wrap of the Dreamweaver carton. For those with more demanding design ambitions, Dreamweaver also has a range of more complex tools which, as one would expect, take more time to master.

Behind the silence of that afternoon in Boston lie a host of vexing and peculiar issues, which have less to do with the technical nature of software design than they do with our fundamental relationship to information technology. Indeed, it is important to stress again that my interests lie not in evaluating Night Kitchen (or any other kind of software) in terms of its merits or faults, but rather to probe the discourses that inform its stated aims and paradigms. As already discussed, one of the more prevalent discursive components is the narrative of emancipation and expressive freedom. Equally pervasive, however, is a form of assumed (or perhaps desired) transparency between our consciousness and the methods and materials we employ to bring them into the material world. The ideal fruit of such a desire is, as Bolter and Grusin describe in *Remediation*, a form of immediacy in which a thought, intention or desire can be instantly manifested into material and temporal reality.

What designers often say they want is an 'interfaceless' interface, in which there will be no recognizable electronic tools—no buttons, windows, scroll bars, or even icons as such. Instead the user will move through the space interacting with the objects 'naturally,' as she does in the physical world. Virtual reality, three dimensional graphics, and graphical interface design are all seeking to make digital technology 'transparent.

(Bolter *et al.*, 1999)

The narrative or discourse of transparency is perhaps one way with which to explain the attitude towards standardization that is reminiscent of Theodor Adorno's condemnation of standardization and its crippling effect on so-called real and authentic creativity and expression (Adorno *et al.*, 1972). In the case of the so-called 'paradox' raised during Stein's presentation, we can draw a parallel between the 'suspension of disbelief' which allows us to ignore the machinery of a mediated experience, such as the stage or the movie screen, and instead be momentarily drawn into the 'world' of the performance. In the case of software such as Night Kitchen, users must suspend their disbelief and adopt, as Zizek puts it, a 'naive trust in the machine' which 'involves a phenomenological attitude, an attitude of trusting the phenomena.' What was perhaps so unsettling for some members of Stein's audience is that, in the course of his presentation, he unwittingly disturbed the trust that Zizek speaks of, thereby re-animating the disbelief that usually keeps the interface and its many rules so well hidden and transparent (Zizek, 1997).

So we come to the matter of trust, or better, faith. In order for a program such as Night Kitchen to become liberating in the sense that Stein describes it, one must suspend suspicions of limited possibilities and instead believe in software's ability to serve as a transparent and unbiased mediator between creative desire and its eventual manifestation. In other words, user-friendly software is based on the illusion that one can really access and control the system in a manner that is entirely free from the constraints and demands of the system itself. Integral to such an illusion is the promise of precise and absolute user control. In addition, in order for software applications to be useful, in the most banal sense of the term, expertise must be attainable in a short period of time, at least in the case of software intended for the consumer market. Usability, to a large degree, depends on the illusion of instant gratification and decreasing limits. If one understands that a particular piece of software is actually a constraint to what could be possible, then the program is a failure. It cannot be used, at least not without a profound level of frustration.

The so-called paradox of user friendly software can thus be characterized by the desire for transparency and a faith that allows one to block out the machines behind and in front of the machine, the codes, parameters and standards that actually allow us to give form to our ideas. Concealing this paradox is the general discourse of emancipation, the myth of freedom that, according to Bolz (writing for the Dutch journal *Mediamatic*) results in a type of giddy or even narcotic form of enslavement that keeps us all relentlessly clicking and surfing.

> In simple terms, user-friendliness means functional simplicity in the face of structural complexity, i.e. easy to operate, but hard to understand. A product's intelligence consists precisely of its ability to conceal this chasm of inscrutability. Use thereby emancipates itself from comprehension. Anyone who still talks about intelligent design now means that a device's use is self-explanatory. Yet this explanation does not lead to understanding but rather to smooth functioning. So to put it stereotypically, user-friendliness is the rhetoric of the technology, which concentrates our ignorance. And this design-specific rhetoric now provides us with the user illusion of the world.
>
> (Bolz, 1998)

Again, we come to the matter of faith. We need to believe that the revolution is digital, that our tools augment instead of restrict possibility and choice. What is problematic, then, is the implied assumption that

ease of use can be directly linked to both individual and collective emancipation and unrestrained creativity. The promise of the computer is thus
grounded partly on the illusion of a one stop form of emancipation where
all needs can be fundamentally met.

3.4   THE QUESTION OF PRACTICE

A set of final questions remains: What does all this mean when it comes
to real and practical application? What is the result of identifying the
apparent paradox inherent in the goals of Night Kitchen? Do such questions lead to the same nervous silence as that which descended upon the
auditorium at MIT when Bob Stein made the request for a better solution
to the one he was currently working on?

One response to such questions relates to an underlying agenda of
this chapter, which is essentially to encourage the idea that software is
a crucial and important cultural text and thus takes its place alongside
other 'texts,' such as literature, music, film and mass media in general.
As cultural critics we must also learn to 'read' software for the purposes
of critique and analysis. Software is not merely a tool but rather a much
more precise expression of aesthetic, cultural, philosophical, ideological
and political (this list could go on) values and paradigms. One consequence here is a blurring of the familiar distinctions between form and
content—a blurring which indicates the need to understand software
more than just a form which contains or carries the content, but as an
element of the content itself.

The first step in working towards such a goal is to identify how software applications are historically, aesthetically, culturally and politically
charged. As such, applications are the fruit of specific ideological vectors, which resonate with the contexts of their time, so to speak. One
familiar and overt example of such vectors is of course Microsoft and
the numerous debates surrounding the company's powerful (and jeal-
ous) grip on the software industry—a grip that could be explored via
discourses of power and economics. In contrast, the firm Linux offers
an example of a company driven by one of the mainstays of the 'hacker
ethic'—information wants to be free. As a leader in the Open Source
Movement, Linux freely distributes the source code for its operating
software, thus allowing its software to be constantly improved upon by

the combined efforts of a world wide community of software engineers. A second and more difficult task is to develop a method for focusing on the 'grammar' or 'language' of specific software domains (such as Windows or Mac OS) in an effort to identify the major tropes, paradigms, codes and 'meaning systems' which make the particular environments 'work.'

Behind such analytic and discursive efforts, however, looms the problematic question of technical expertise. Thus, in order to discuss the semiotics of Director or the ideologies implicit in the Mac OS, for instance, is it necessary for me to be fluent in the programs themselves? Or is it enough to be aware of the discursive environment of the total application, which includes advertising campaigns, owners' manuals, and press releases along with the general aesthetics of the application interface? This is a tricky question, indeed. In the case of Director, the successful user must master a programming language known as 'Lingo' in addition to being able to navigate the overall interface. Now, certainly it could be argued that 'Lingo' carries some ideological weight and thus the critic would be required to learn it if s/he has any intention of analysing it from a cultural perspective.

Such concerns are not entirely new and are related, in fact, to the discipline of science studies, which also deals with the issue of how much scientific knowledge is necessary to engage in a cultural critique of science. As Andrew (among others) discusses, the standard argument that only 'real' scientists have the credentials necessary to engage in any assessment of science works to deflate its critical assessment. In other words, if you can't even work out a differential equation, then your voice as a critic is pretty much irrelevant. Naturally, there is much to dispute here, not the least of which is the fact that such forms of 'technocratic credentialism' are primarily an exercise in power and are thereby rooted in the mundane worlds of economics, ideology and politics (Andrew, 1999). The 'critical study' of software applications is confounded with similar complications. If software is indeed to be studied as a 'cultural text,' then to some degree it will be necessary to penetrate the interface in order to better understand the total nature of a particular software environment. Again, this comes back to the point that software is more than just a tool, more than just a carrier of content. Consequently, it is necessary to pay attention to its deeper structures—a task that would seem to benefit from a combination of theoretical and technical expertise.

Such a combination is perhaps one way to disrupt the 'user-illusion' identified by Bolz (1998), to disengage the 'delusion of familiarity' and the empty promise of trouble-free emancipation. It is perhaps also a way in which to overcome the awkward silence between good intentions and their eventual and often compromised implementation.

3.5 REFERENCES

Adorno, T. and M. Horkheimer (1972) *Dialectics of Enlightenment* (transl. J. Cumming), The Seabury Press, New York.

Andrew, R. (1999) 'The Challenge of Science', in S. During (Ed.) (1999) *The Cultural Studies Reader*, London, England, pp. 303–304.

Apple Home Page (1999) <http://www.apple.com/imac/>, December 9, 1999.

Bolter, J. D. (1991) *Writing Space: The Computer, Hypertext and this History of Writing, L. Erlbaum Associates,* New Jersey.

Bolter, J. D. and R. Grusin (1999) *Remediation: Understanding New Media*, MIT Press, Cambridge, Massachusetts.

Bolz, N. (1998) 'The Use-illusion of the World: On the Meaning of Design for the Economy and Society in the Computer Age', *Mediamatic*, 9 (2/3), 61.

Cohen, A. (1999) 'The Attic of E', *Time*, 27 December, 1999.

Dery, M. (1996) *Escape Velocity: Cyberculture at the End of the Century*, Grove Press, New York.

Eastgate Systems home page (1999) <http:www.eastgate.com>, December 10, 1999.

Landow, G.(Ed.) (1994) *Hypertext/theory*, Johns Hopkins University Press, Baltimore.

Landow, G. (1999) 'Hypertext as Collage Writing', in P. Lunenfeld (Ed.) (1999) *The Digital Dialectic: New Essays on New Media*, MIT Press, Massachusetts.

Plant, S. (1997) *Zeroes and Ones: Digital Women and the New Techno-culture*, Doubleday, New York.

Taylor, M. (1999) *About Religion: Economics of Faith in Virtual Culture*, University of Chicago Press, Chicago.

Turkle, S. (1995) *Life on the Screen: Identity in the Age of the Internet*, Simon and Schuster, New York.

Zizek, S. (1997) *The Plague of the Fantasies*, Verso, New York.

4

Business Modelling as a Foundation for Multimedia Development— Concerning Strategic, Process and Systems Levels in Organizations

Anders G. Nilsson

Information Systems, Karlstad University

4.1 BUSINESS MODELLING—SOME NEEDS IN PRACTICE

Business managers and systems experts often experience frustration when information support for business operations is being discussed.

Perspectives on Multimedia R. Burnett, Anna Brunstrom and Anders G. Nilsson
© 2004 John Wiley & Sons, Ltd ISBN: 0-470-86863-5

A major part of the problem is due to lack of essential communication and accusation from both sides is often the result. 'The business people do not know what they want!' or 'The systems people do not understand what we need!' However, communication is more than talking. It requires a firm base for common understanding, a common language.

It is in this sense a specific method for business modelling can be a useful tool to create a common language between business and systems people (Nilsson *et al.*, 1999). By a 'method', we mean concrete guidelines or prescriptions for a systematic way of working with development tasks in organizations. It is possible to distinguish three main constituents of a method (Nilsson, 1995):

- *Perspectives*: basic principles, views and assumptions that influence the proposed work between affected interest groups in business development.

- *Work tasks*: guidelines to manage different issues and decisions during the development process. The development work is divided into a number of perceivable and delimited tasks, for example, creation of different types of business models.

- *Interest groups*: list of actors who could participate during the development process, together with possible forms of collaboration. Specify 'who is responsible for what' during the development work.

Successful management and planning of business enterprises require rich and precise models (descriptions) of both the real business operations and their links to the supporting information systems. Business modelling is a way of working to achieve such descriptions. Today, we have many established methods and models for development work in the market (Avison and Fitzgerald, 2003).

There has been much debate over the years about the actual effects of method use for business modelling in practice. Below we summarise some needs for using methods to support development work:

- *Requirements specifications*: the need to produce exact, consistent and complete requirement specifications for designing the future business and information operations.

- *Explain IT possibilities*: the need for explaining how new IT possibilities can improve business processes and sharpen corporate strategies in organizations.

- *Describing business flows*: the need for describing and coordinating the complex nature of material flows, information flows and cash flows in enterprises.

In today's business world, information support has become a more integrated part of business operations and, in many cases, a vital part of the business mission itself. In fact, information systems can also create new business opportunities for companies to reinforce their competitive edge in the market place.

In most cases development of corporate strategies, business operations and its information support, are often carried out as separate change processes and as independent projects in companies. There is very limited, or no, organizational coordination and timing between business and systems development. The challenge here is to use efficient methods to make a progress in reaching a more holistic view of development work in organizations.

4.2 BUSINESS MODELLING—THREE LEVELS OF DEVELOPMENT WORK

A general view in business administration research is that products are going through a life cycle. In the same way we can regard business and information operations from a life cycle perspective (Nilsson, 1995). A proposal for such a life cycle could be as follows:

(1) business development;

(2) operation and maintenance management;

(3) business termination.

We will here focus on the first stage of the life cycle, namely business development. Business modelling is an essential part of business development. By business modelling we refer to how different actors or people are using models and methods in order to understand and change business processes, together with their information systems, in companies and organizations. Business development generally consists of different tasks (Österle, 1995; Tolis and Nilsson, 1996; Fitzsimmons and Fitzsimmons, 1998; Nilsson, 1999). We can recognise three levels of development work in practice with a distinct scope and focus (see Figure 4.1):

- *Strategy development*: focusing on corporate strategies for improving the relations between our company and the actors in the market environment, e.g. customers, clients, suppliers and business partners (cf. Ansoff and McDonnell, 1990; Kaplan and Norton, 1996; Porter, 1980, 1985).

- *Process development*: focusing on how to make the business processes more efficient within our company. The workflow between different functions in an organization is designed in a new and better way (cf. Davenport, 1993; Edvardsson *et al.*, 2000; Ishikawa, 1985; Jacobson *et al.*, 1994; Rummler and Brache, 1995).

- *Systems development*: focusing on how support from information systems (IS) and information technology (IT) can be useful resources and efficient enablers for running business operations more professionally and strengthen the competitive edge of our business achievements (cf. Avison and Fitzgerald, 2003; Booch *et al.*, 1999; Fitzgerald *et al.*, 2002; Hawryszkiewycz, 2001; Jacobson *et al.*, 1993; Lundeberg *et al.*, 1981; Nilsson and Pettersson, 2001; Yourdon, 1993).

Today no 'super' method giving support to the whole development work in a company is to be found in the market. On the contrary, the current methods for business modelling are delimited to a concrete level of development work according to Figure 4.1. Furthermore, methods for the same level somewhat 'attack' different problems or perspectives in business modelling.

Figure 4.1 Three levels of development work in business modelling.

There is not a need to work in a 'top-down' fashion from strategy development through process development down to systems development. In a real case, one can start at a certain development level and let the outcome of this work trigger some other levels upwards and downwards, often in several rounds (indicated by the arrows in Figure 4.1). We can therefore regard the development levels as essential inquiry areas during a whole change process.

From earlier research it is a well-known fact that there are real communication gaps between top managers, business leaders and systems experts when developing organizations (Nilsson *et al.*, 1999; Penker and Eriksson, 2000). It is therefore useful to distinguish three levels of development work, namely strategy development, process development and systems development.

4.3   BUSINESS MODELLING AND MULTIMEDIA

Business development means a change work in general for organizations working with all three levels of strategy, process and systems development. Multimedia development is a special kind of change work focusing on a smaller, more concrete and narrow product for digital media in companies (cf. Keen, 1997). Multimedia can be used in different situations, settings or contexts. The main characteristic is that a multimedia development process has to deal with a high degree of interactivity between the user and the system (see, for example, Chapman and Chapman, pp. 13–15).

Figure 4.2 shows that a multimedia solution can be used as a tool for working out corporate strategies, business processes and IS/IT supports. This means that 'multimedia' could come into contact with all

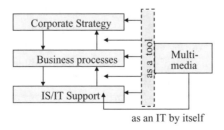

Figure 4.2 Multimedia as a tool for development work.

three development levels. There is also a good possibility of using multimedia as an mediating link between two of the development levels. Furthermore, the multimedia technique can be used as an IT instrument by itself for systems development. In total, we have six cases of how business modelling and multimedia can affect each other; concerning strategy, process and systems levels in organizations.

The purpose of business modelling is to bridge the communication gap between actors and create a good coordination between the three development levels in organizations. Development of multimedia products is a rather new form of change work in today's organizations. Multimedia development can affect key issues on strategic, process and systems levels, and it is therefore interesting to study the role of digital media when developing future organizations. Business modelling constitutes, in this respect, a foundation for multimedia development as it has always been when working for other kinds of development work in organizations, e.g. in market development, manufacturing development, product development and service development.

4.4   BUSINESS MODELLING—SYSTEMATIC WORK IN PHASES

Business modelling is a concrete instrument for working with strategy development, process development and systems development in organizations. It supports a systematic way of working with different issues on all three development levels in a company. Methods for development work often propose a set of concrete phases in a life cycle manner, starting from a problem situation and ending up with some sort of business solution (cf. Davenport, 1993; Lundeberg *et al.*, 1981; Jacobson *et al.*, 1993; Porter, 1980, 1985; Yourdon, 1993).

Figure 4.3 (left-hand side) shows a comprehensive life cycle approach for business modelling with systematic work in four phases, as is general for strategy, process and systems development in organizations. The development work starts with a change analysis where we identify and penetrate different user problems and describe our business engagements in various enterprise models. In the next phase, we utilize the enterprise models as a base or platform to formulate a requirements specification from a user point of view. From this rather detailed specification, we implement some kind of business solution like a corporate strategy,

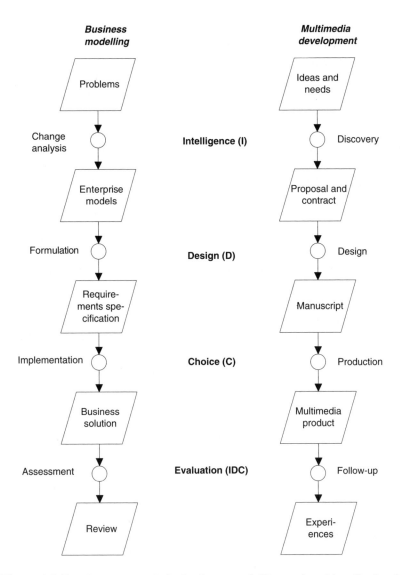

Figure 4.3 Development work for business modelling and multimedia development.

business process and/or information system. On a regular basis we assess the business solution from different aspects and compiling the results in a review document.

The process behind multimedia development can be described in a similar manner with four concrete phases for producing a digital media product (cf. Chapman and Chapman, 2000; Dixon, 1998; Earl and Khan, 2001; Elin, 2001; Lopuck, 1996). Figure 4.3 (right-hand side) shows a possible life cycle approach for multimedia development. The development work starts with discovery of user ideas and needs. This is a very creative phase and ends after a while up with a proposal for design together with a contract. The aim of the design phase is to work out a concrete manuscript that describes the content and structure of the media solution. Based on this manuscript, we accomplish the production phase with a multimedia product for use as an output. We should do regular follow-ups of how different users perceive and experience the multimedia solution in their daily work.

The described life cycles of business modelling and multimedia development have much in common. This can be explained from the IDC approach for general decision making worked out in the mid 1960s by the Nobel prize winner in economics (1978) Herbert Simon. Simon (1965) states that all kinds of decision making are going through three phases: Intelligence (I), Design (D) and Choice (C).

> The first phase of the decision-making process—searching the environment for conditions calling for decision—I shall call *intelligence* activity (borrowing the military meaning of intelligence). The second phase—inventing, developing, and analysing possible courses of action—I shall call *design* activity. The third phase—selecting a particular course of action from those available—I shall call *choice* activity.... It may be objected that I have ignored the task of carrying out decisions. I shall merely observe by the way that seeing that decisions are executed is again decision-making activity.
>
> (Simon, 1965, pp. 54–56)

Development work is a kind of decision-making activity as described by Simon (1965). The life-cycle approaches for business modelling and multimedia development in Figure 4.3 follow the IDC pattern. Intelligence corresponds to change analysis in business modelling and discovery in multimedia development. Design is what is called formulation in business modelling and just what is called design in multimedia development. Choice is made through an implementation activity in business modelling and production phase in multimedia development. The evaluation phase, called assessment in business development and followup in multimedia development, can be characterized as a new decision-making activity with its own IDC triple according to Simon's view.

This means that life cycle approaches for business modelling and multimedia development are scientifically anchored and, therefore, are also in harmony with each other. Business modelling can be used as a base or platform for multimedia development, but lessons learned from work with multimedia development can, on the other hand, be used to enhance the qualities of business modelling in practice.

4.5 BUSINESS MODELLING OF VALUES, OPERATIONS AND OBJECTS

The requirements specification is an instrument for accurate descriptions of contents, structure and contributions that a desired business solution would have in an organization. The specification should illustrate different users' demands on the new business situation. In business modelling the requirements specification is the key document for realizing and implementing the wished-for business concepts into a useful and purposeful action in our organization. Corresponding to the requirements specification document is the manuscript in multimedia contexts (see Figure 4.3). This manuscript has a key role for the production of multimedia arts in the same way that the requirements specification has for business modelling of strategy, process and systems solutions.

A current problem with requirements specifications is that they are treated in an one-dimensional way, as they illuminate only a limited perspective of business modelling. In change work we need to describe business achievements from many different perspectives (cf. Olle *et al.*, 1991; Penker and Eriksson, 2000; Nilsson, 1995; Nilsson *et al.*, 1999), such as (Figure 4.4):

- *Values*: concerning goals, intentions, visions, problems, strengths, strategies, actors, force fields, influences, critical success factors, casual relations, etc.

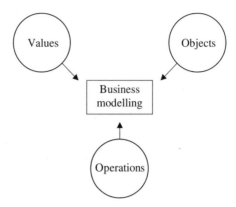

Figure 4.4 Business modelling from three perspectives.

- *Operations*: concerning functions, activities, processes, behaviours, events, tasks, rules, routines, transformations, work flows, etc.

- *Objects*: concerning data, concepts, categories, information, connections, contents, areas of responsibility, standard components, etc.

In the development of organizations we need to turn the descriptions or models of the business situation over in our minds. By changing perspective and observing our achievements from many different angles, we will gain a deeper understanding of the underlying mechanisms within the organization. Thus we will have a more solid basis from which to suggest vigorous changes of the daily work. Business modelling can be seen as an act of modelling different values, operations and objects in our organizational contexts. This kind of multidimensional modelling could be a good base for conducting more efficient multimedia development in our organizations. The values perspective emphasizes the aim or purpose behind a specific multimedia product. The operations perspective accentuates a clever use of multimedia crafts in the users' daily work situation. The objects perspective stresses the importance of a good content of the multimedia solution in a given business context.

Lessons learned from business modelling are that a multiview approach to development work will give sustainable results and good business performance representing valuable knowledge to users and developers of multimedia products. Modelling of values, operations and objects behind investments in multimedia concepts will give more transparency and clarity in our use of digital media to support different kinds of business activity in industry and public services.

4.6 BUSINESS MODELLING FOR MULTIMEDIA—MAIN MESSAGES

Business modelling concerns development work in organizations, meaning that people are using models and methods in order to understand and change their business situation. Multimedia development is a special kind of change work, with a strong focus on producing a digital media solution to support some specific part in a business context. What have business modelling and multimedia development in common? This has been the main research question behind this contribution. The overall conclusion is that there is good potential in connecting business modelling and multimedia development to each other in a productive manner. Therefore, business modelling could be a valuable foundation for more efficient creation and production of multimedia products in practice. We will finish with the main messages regarding the interplay between business modelling and multimedia development:

- Business modelling comprises development work on strategy, process and systems levels in organizations. When appropriate, use the multimedia technique as a tool to make the development work more effective through all these levels.

- Business modelling means a systematic way of working with development issues in organizations. Investment in multimedia products is often a really complex undertaking or engagement in today's organizations. Therefore, use a specific method for systematic work in concrete phases towards the final multimedia product.

- Business modelling has, as its principal lodestar, to penetrate a business problem from different angles or perspectives. If possible, try to model values, operations and objects in requirements specifications and manuscripts in order to get a valuable multiview of the complexities faced with multimedia development.

4.7 REFERENCES

Ansoff, I. and E. McDonnell (1990) *Implanting Strategic Management,* Second Edition, Prentice-Hall, New York.

Avison, D.E. and G. Fitzgerald (2003) *Information Systems Development: Methodologies, Techniques and Tools,* Third Edition, McGraw-Hill, London.

Booch, G., J. Rumbaugh and I. Jacobson (1999) *The Unified Modelling Language User Guide,* Addison-Wesley, Reading, Massachusetts.

Chapman, N. and J. Chapman (2000) *Digital Multimedia,* John Wiley & Sons, Inc., New York.

Davenport, T.H. (1993) *Process Innovation: Reengineering Work through Information Technology,* Harvard Business School Press, Boston, Massachusetts.

Dixon, S. (Ed) (1998) *Multimedia Pathways: A Development Methodology for Interactive Multimedia and Online Products for Education and Training,* Impart Corporation Pty Ltd, Brisbane, Australia.

Earl, M. and B. Khan (2001) 'E-Commerce Is Changing the Face of IT', *MIT Sloan Management Review,* **43**(1) 64–72.

Edvardsson, B., A. Gustafsson, M.D. Johnson and B. Sandén (2000) *New Service Development and Innovation in the New Economy,* Studentlitteratur, Lund.

Elin, L. (2001) *Designing and Developing Multimedia: A Practical Guide for the Producer, Director and Writer,* Allyn and Bacon, Boston.

Fitzgerald, B., N.L. Russo and E. Stolterman (2002) *Information Systems Development: Methods in Action,* McGraw-Hill, London.

Fitzsimmons, J.A. and M.J. Fitzsimmons (1998) *Service Management: Operations, Strategy, and Information Technology,* Second Edition, McGraw-Hill, Boston, Massachusetts.

Hawryszkiewycz, I. (2001) *Systems Analysis and Design,* Fifth Edition, Prentice Hall/Pearson Education Australia, Sydney.

Ishikawa, K. (1985) *What is Total Quality Control? The Japanese Way*, Prentice-Hall, Englewood Cliffs, New Jersey.

Jacobson, I., M. Christerson, P. Jonsson and G. Övergaard (1993) *Object-Oriented Software Engineering: A Use Case Driven Approach*, Fourth revised printing, Addison-Wesley, Wokingham, England.

Jacobson, I., M. Ericsson and A. Jacobson (1994) *The Object Advantage: Business Process Reengineering with Object Technology*, Addison-Wesley, Wokingham, England.

Kaplan, R.S. and D.P. Norton (1996) *The Balanced Scorecard: Translating Strategy into Action*, Harvard Business School Press, Boston, Massachusetts.

Keen, P.G.W. (1997) *Business Multimedia Explained: A Manager's Guide to Key Terms and Concepts*, Harvard Business School Press, Boston, Massachusetts.

Lopuck, L. (1996) *Designing Multimedia: A Visual Guide to Multimedia and On-line Graphic Design*, Peachpit Press, Berkeley.

Lundeberg, M., G. Goldkuhl and A.G. Nilsson (1981) *Information Systems Development: A Systematic Approach*, Prentice-Hall, Englewood Cliffs, New Jersey.

Nilsson, A.G. (1995) 'Evolution of Methodologies for Information Systems Work: A Historical Perspective', in Wrycza, S. and J. Zupancic (Eds) *Proceedings of The Fifth International Conference on Information Systems Development—ISD'96*, University of Gdansk, Poland, pp. 91–119.

Nilsson, A.G. (1999) 'The Business Developer's Toolbox—Chains and Alliances between Established Methods', in Nilsson, A.G., C. Tolis and C. Nellborn (Eds) *Perspectives on Business Modelling: Understanding and Changing Organisations*, Springer, Berlin, pp. 217–241.

Nilsson, A.G. and J.S. Pettersson (Eds) (2001) *On Methods for Systems Development in Professional Organisations: The Karlstad University Approach to Information Systems and its Role in Society*, Studentlitteratur, Lund.

Nilsson, A.G., C. Tolis and C. Nellborn (Eds) (1999) *Perspectives on Business Modelling: Understanding and Changing Organisations*, Springer, Berlin.

Olle, T.W., J. Hagelstein, I.G. Macdonald, C. Rolland, H.G. Sol, F.J.M. Van Assche and A.A. Verrijn-Stuart (1991) *Information Systems Methodologies: A Framework for Understanding*, Second Edition, Addison-Wesley, Wokingham, England.

Österle, H. (1995) *Business in the Information Age: Heading for New Processes*, Springer, Berlin.

Penker, M. and H.-E. Eriksson (2000) *Business Modeling with UML: Business Patterns at Work*, John Wiley & Sons, Inc., New York.

Porter, M.E. (1980) *Competitive Strategy: Techniques for Analyzing Industries and Competitors*, The Free Press, New York.

Porter, M.E. (1985) *Competitive Advantage: Creating and Sustaining Superior Performance*, The Free Press, New York.

Rummler, G.A. and A.P. Brache (1995) *Improving Performance: How to Manage the White Space on the Organization Chart*, Second Edition, Jossey-Bass Publishers, San Francisco.

Simon, H.A. (1965) *The Shape of Automation for Men and Management*, Harper and Row, New York.

Tolis, C. and A.G. Nilsson (1996) 'Using Business Models in Process Orientation', in Lundeberg, M. and B. Sundgren (Eds) *Advancing Your Business: People and Information Systems in Concert*, Chapter VIII, The Economic Research Institute (EFI), Stockholm School of Economics, Stockholm. Also <http://www.hhs.se/im/efi/ayb.htm>.

Yourdon, E. (1993) *Yourdon Systems Method: Model-Driven Systems Development*, Yourdon Press, Prentice-Hall, Englewood Cliffs, New Jersey.

5

How Should Interactive Media Be Discussed For Successful Requirements Engineering?

Lennart Molin and John Sören Pettersson
Information Systems, Karlstad University

5.1 SPECIFYING REQUIREMENTS

Specifying requirements for a multimedia system is difficult and is often carried out in an informal way, as data presented in this chapter shows. Requirements engineering in traditional information systems development is, on the other hand, often done in a more formal and thorough way, as discussed by Nilsson in this book (Chapter 4). It is easy to give arguments that the requirements specification ought to be explicit and thoroughly documented (see also Davis, 1993, p. 181, for an enumeration of criteria of a 'perfect software requirements specification'):

Perspectives on Multimedia R. Burnett, Anna Brunstrom and Anders G. Nilsson
© 2004 John Wiley & Sons, Ltd ISBN: 0-470-86863-5

(1) The requirements specification is the basis for a contractual agreement between supplier and customer.

(2) The requirements specification is also the foundation for design and construction.

(3) The requirements specification is needed to verify the completed system.

In this chapter we contrast common practice in multimedia development against this notion of the requirements specification.

Multimedia systems are characterized by the important role the systems' extrovert parts have. That is simply the meaning of 'media' in 'multimedia'. Such systems are, to a large extent, defined through their user interfaces. This is not to deny that there are internal properties of such systems as well, but such properties do not necessarily have to surface in the requirements specifications when the essential part of the systems is their user interface. Similarly, most information systems nowadays have their 'multimedia' parts, that is, some user interfaces that rely on today's user interface standards. Our discussion is to some extent applicable also to these parts of such systems. It could be noted that some authors, like Löwgren (1995), make a case for separating external from internal software design.

We present results from an interview survey on multimedia systems development, and furthermore present a proposal for multimedia 'articulation', in particular interactive media articulation, in requirements engineering. The question addressed is if and how it is possible to gather and write requirements specifications in a way better suited to multimedia development. To compare, the architects make drawings to specify constructions that will be spatial (houses and their ambient environment). That is, they make representations based on spatial relations to represent constructions that will be spatial. One may ask to what extent multimedia systems specification could be done multimedially. In particular, we address the questions of how the interactive aspects of multimedia systems could be specified without a media translation. We base the discussion on a computer system developed at Karlstad University, a system which could be used to remedy some of the problems prevailing in multimedia development work.

To acknowledge that some of the answers are in no way unique to multimedia development, we note that many researchers discuss specification of software systems. Tannenbaum (1998) relates critiques by Blum (1996) and Isaacs *et al.* (1995) towards traditional models such as the waterfall model and the spiral model. They argue that computer scientists, with the roots of their science in mathematics and engineering, concentrate on the *formal specification of precise solutions* in hardware and software (Tannenbaum, 1998, p. 469). Blum (1996, pp. 255–258) argues that such an approach may be appropriate for bridges but not for complex software. He further argues that the development process should be user oriented and should therefore be led by the user's needs. Developers should use an approach that he calls 'adaptive design', which allows for constant revaluation of the problem at hand and the needs of the users. Such an approach will allow cooperation and compromise, which will lead to increased productivity and satisfaction among users. Bubenko and Kirikova (1999, p. 244) note that the 'working languages' of different stakeholders may differ significantly, and that flexible forms of representation of the content of the requirements specification will be necessary. Similarly, Löwgren notes that concepts developed in the design phase can 'serve as the common ground necessary for communication, provided that they are expressed in media that do not discriminate the users' (Löwgren, 1995, p. 93).

This discussion is interesting in the light of the properties that characterize actual multimedia development. The interview survey showed that a very detailed and explicit requirements specification, in the sense that is usual in traditional systems development projects, was rarely constructed. Two prominent causes appear from the data and will be discussed in subsequent sections:

- few of those involved in developing multimedia have a tradition of working with an explicit requirements specification (Section 5.2);
- it is difficult to produce a detailed requirements specification in multimedia development (Section 5.4).

In relation to the last bullet, it is interesting to note how Blum (1996) describes the different ways of categorizing requirements. One way, which relates to the development of multimedia in particular, is to divide the requirements into closed and open. Closed requirements are well defined and stable and can be described formally with some form of existing notation. Open requirements, on the other hand, are dynamic and difficult to understand. There is great uncertainty as to whether the specified product can meet expectations as regards these requirements, especially since they tend to change as the user's understanding of the system deepens. Multimedia products normally contain a large number of open requirements.

The problems identified in the interview survey are listed in Section 5.4 with discussions of the underlying causes. For each problem, the section also includes a discussion of the prospect of remedying the problems with a computer-based support (introduced in Section 5.3) for visualizing possible interactive designs. In this way, the section proposes an answer to the question as to how interactive multimedia should be discussed. Section 5.4 ends with an analysis of how open properties relate to the demand for explicit and assessable requirements implied by the three arguments listed in the beginning of this chapter.

Section 5.5 compares two studies from other areas where written texts may not be the most natural medium for discussion and specification. This comparison is made in order to indicate problems that may be encountered in a multimedia-based specification for a multimedia system.

Finally, in Section 5.6, we present how even non-designers could perform requirements work with our computer-based visualizer. Naturally, this tool may be used also by content professionals for systems design. However, here we extend the question of how to gather requirements information from various user groups to a question of whether such a group could deliver a requirements specification. At the heart of the discussion lies the communication between developers and customers/users.

5.2   REQUIREMENTS WORK IN MULTIMEDIA SYSTEMS DEVELOPMENT

5.2.1   An interview survey of multimedia developers

Multimedia developers draw their background from many different fields (Tannenbaum, 1998). Only a few of them have education or work experience in traditional software development. For those who draw their background from other areas, the concept of requirements specification is probably less familiar. This does not mean that the actual content of requirements work is unknown to them. In many different types of development work, there is a phase that circles around similar problems and leads to results which resemble the requirements specification.

The interviews mentioned in this chapter are part of a survey that is given extensive treatment elsewhere (Molin, 2002). Nine interviews were done in 1999, eight interviewees were multimedia producers and one was a systems analyst working mostly with non-multimedia systems. A semi-structured questionnaire was used which covered not only requirements specification but also a broad range of other topics, such as the company's organization, user involvement, use of methodologies and prototypes, creativity, quality measures, production tools and documents, information and communication aspects of the development process, the customer's role and relationship to the development team, and factors for success or failure. The respondents are referred to as R1–R9.

5.2.2   One result of the survey: weak tradition concerning requirements work

The multimedia producers who were interviewed generally used a small number of clearly specified written documents, seldom more than four or five different types. A couple of companies keep the requirements specification 'in the head'. Three companies use a template from the industrial association Promise (Producers of Interactive Media in Sweden, 1998).

Below are a number of quotes illustrating the use of requirements specifications. The first four clearly show that the documentation of the process of specifying the requirements is not particularly extensive.

> It's not very detailed when it comes to the product but all the steps are included. All the overarching headings and the like and something about what we were thinking of doing. But then it's something that takes shape as we write the script, it's not particularly detailed.
>
> (R1)

> The most important thing is speed. We have done a fantastic amount in three weeks. We don't have a requirements specification, or a script but we know what

we want. We shall have to see how far we can get in two months with the resources we have. That's not strictly true, we have a requirements specification with the functions we want. Our resources are limited and this determines what we can do. We have priorities 1, 2 and 3. At a given time it must be good enough to launch. We identify the most important modules. There will never be a detailed script, those developing the system must have a feeling for what they're doing. It's a never-ending development process.

(R4)

Attitudes to the requirements specification vary considerably, as the following quotes show:

No, we never have any.

(R9, on requirements specifications as documents)

The requirements specification is important but it doesn't exist as a set of papers, it's continually being changed, refined.

(R4)

In the next quote the respondent expresses her appreciation of the documentation of specifications:

We follow the Promise specification template. It works quite well, covers a lot of things. It is applied in connection with the pilot study. It forms the basis for the next phase. It's quite new. Promise took the initiative. Customers often think the agreement is too biased towards the producer. Good to have it as a starting point and then diverge from it, better than not having anything and forgetting. It was much worse before Promise.

(R5)

She goes on to provide good arguments for spending time on the requirements specifications:

If you don't do the preparatory work properly, the results are catastrophic; you must have some form of pilot study or script. If you start production and then have to make changes, it becomes very expensive. You must be in full agreement with the customer about what they want, what exactly is to happen, what is to be there. Otherwise it will be expensive, take a long time and not be particularly good. If you don't draw up a requirements specification, you won't have the prerequisites that different people need. You simply must know what the requirements on the product are.

(R5)

Similar thoughts are revealed in the next quote:

The script is part of the requirements specification, it emerges from the early stages of the pilot study. The first part of the specification is in the form of answers to early questions: target group, technical matters, how, where and when, maintenance. The script is the second part. The first part must be documented before the script is prepared. You don't document it afterwards. Promise provides a template.

(R2)

The picture that emerges from the study of the process of specifying the requirements reveals that the routines for this process are not strictly formalized. On the other hand, some of the respondents have ideas and wishes for improving routines and structures. The templates that have been devised by the industrial association have had a positive reception from those who have used them.

The survey reported above shows, furthermore, that there are many different kinds of factor that constitute obstacles for the work of creating detailed requirements specifications in multimedia systems development. In Section 5.4 we will describe these factors and also how interface prototyping could aid the process. Immediately below we describe an interface prototyping tool, which we call Ozlab.

5.3   OZLAB—A SIMPLE TOOL FOR PROTOTYPING INTERACTIVITY

The Ozlab software was developed at Karlstad University in order to make it easy to test, at a conceptual stage, the interactivity of multimedia products. In particular, it was necessary to facilitate testing of interactivity with respect to the graphical user interface, GUI. The term GUI as used here does not mean simply drop-down menus and dialogue boxes but more graphically oriented interaction. Although the GUI has existed in commercially available systems since the introduction of Lisa and Macintosh by Apple Computers in the 1980s, graphical interaction is not very much exploited except in multimedia games and in multimedia for children.

There is an experimental technique often employed in language technology called 'Wizard of Oz' (Dahlbäck *et al.* 1993). In Wizard-of-Oz experiments, a test person thinks he is writing or speaking to the computer in front of him when in fact the test manager is sitting in the next room interpreting the user's commands and providing appropriate responses (see Figure 5.1). In this way, natural language processing of either text or speech can be simulated even when there is no unit available that understands natural human language. One major advantage is that dialogue structure can be tested before one decides on how clever the automatic interpretation has to be (e.g. whether to analyse individual words only or also syntactic structures). Automatic interpretation of text or speech is difficult and the Wizard-of-Oz technique thus gives systems developers a chance to test systems before it is even possible to make them. This kind of manual prototype could act as a stand-in for working prototypes.

Figure 5.1 In Wizard-of-Oz experiments, the test manager sits in the next room interpreting the user's commands and provides the system's responses. The system functionality is thus faked in a way that makes it possible to test system proposals without programming prototypes. (Illustration by Anders Karlsson, 2001. © Karlstad University)

However, it seems as if the mere deceptive properties of this technique have not been fully exploited. Since the system looks real to the test user, one could use Wizard-of-Oz mock-ups to test design ideas when there are reasons to believe that simple tests by sketches and slides, as preferred and recommended by usability experts (Klee, 2000, among others), will not provide the right responses. This could be the case when testing concepts on children or on adults with severe learning difficulties, or when the look-and-feel is to be tested. The latter point is especially important when the person (company) who orders the system is inexperienced and needs a means of seeing various variants of his own suggestions in working practice, that is, he needs to see the system proposals in interaction with users.

In our work, we take it for granted that simple paper prototypes are in some instances not suitable for testing interactive software. At the same time, it is quite expensive to make a fully fledged product just to see if it works. Consequently, a plain GUI with only 'icons', background, and 'drag-and-drop' functionality could be useful if the experimenter could mimic the behaviour of a planned multimedia product, as in the Wizard-of-Oz experiments conducted in projects on natural language processing.

In fact, some experiments and system developments have been conducted which are described as 'multimodal', often implying that the test user has access to several input channels (like voice, typing, mouse, touch screens, sometimes even eye direction detectors). For instance, in their paper 'Multimodality from the User and System Perspective', Coutaz *et al.* (1995) describe NEIMO, a Wizard-of-Oz system designed to care for multiple types of input (see also Caelen, 1999, and Caelen and Millien, 2002, about MultiCom). For our Ozlab we have focused very much on the ordinary PC set-up. Primarily, Ozlab is intended for the *interaction* to be simulated. It does support showing video clips on the test user's monitor, but a 'multimedia' piece like that is not really dependent on interaction with the user. Therefore, we have concentrated on how to let the wizard respond graphically by direct manipulation of the user's GUI. Furthermore, adding voice to what happens on the screen, as is common in multimedia products, is easily done in Wizard-of-Oz experiments. Other sounds have to be pre-recorded just like video clips, but can be activated on the Wizard's command. This allows for user testing before the difficult interactive behaviours have been programmed. Thus, our system supports the imitating of many ordinary multimedia concepts but is especially designed for experimenting with crucial interactive aspects.

A note on media types could be appropriate. The graphical output does not demand the same exactness as text output in ordinary Wizard-of-Oz prototyping where spelling mistakes will be very revealing. This fact makes it easier to test graphical 'dialogues' than to test purely linguistic dialogues. Pre-written sentences can easily be made visible in Ozlab like any other graphical object. Naturally, Ozlab allows for free text output as well, and also free voice responses through a voice disguising unit, but is not primarily designed to support exact linguistic output in contrast to the above-referred study by Dahlbäck *et al.* (1993).

It should be noted that because not much more than the graphics and a few wizard-supporting functions attached to each graphic object are needed before test trials can be run, Ozlab could be used for improvisation—explorative experiments—as well as for ordinary tests of a pre-defined response scheme (for an extensive account, see Pettersson, 2003).

In what follows, it will be demonstrated how a Wizard-of-Oz supporting system like Ozlab can aid the process of formulating the requirements specification for multimedia systems. It should perhaps be pointed out, that the Ozlab software could be run outside the laboratory on any pair of laptops or ordinary PCs, which makes it possible to carry out mock-up prototyping on site.

5.4   CHALLENGES FOR MULTIMEDIA REQUIREMENTS WORK

As already mentioned, the interview survey showed that there are many factors complicating the multimedia requirements work. In this section, we will describe these factors but also indicate the work in progress with Ozlab, namely how mock-up prototyping with apparently functioning interaction routines could aid the process.

5.4.1   Choosing among different media

Being able to choose between different media is important. Does the customer have enough knowledge to make the best decisions? From the interview survey, it is obvious that for many a customer it is difficult to specify such requirements without guidance. The form of guidance could, however, be discussed. Educating the customer rather than showing only one tentative solution will be crucial. The education may, or probably will, have to include showing alternatives. This will enable the customer to choose and even infer other possible solutions. Such work will naturally be costly if it has to be performed by programming prototypes. In Ozlab simple multimedia sketches, including interactivity, can be shown and demonstrated, and customers/users are able take an active part in making layouts. This is also connected to the form of interaction (see next Section 5.4.2).

However, the benefits of customer and end-user involvement should not obscure the fact that expert knowledge is needed. Which media are chosen entails technical and economical consequences:

> It's difficult to measure how large a production is. If the customer says 60 images, that perhaps means 30 images for us and an animation that takes a long time. What the customer thinks is little, is often a lot for us. We have to explain what takes time. We are getting better and better.
>
> (R2)

It is very difficult for the customer to have the necessary insights about these consequences. This part of the definition of the system requirements demands that both parties engage wholeheartedly in a mutual discussion in order to reach a satisfactory result.

5.4.2 Multimedia entails using different media interactively

Interactivity is one of the cornerstones of any multimedia system. It is what distinguishes a multimedia system from a mere digitized videofilm. Automated interactivity is a relatively new phenomenon and knowledge is limited. To this observation it should be added that the concept of 'interactivity' has many faces. Thus speaks Tannenbaum about interactivity:

> ... it is a complex and multifaceted concept that scholars approach in many different ways, and it is a crucial aspect of multimedia that should be given primary consideration in its development and analysis. There is general agreement that interactivity enhances multimedia but that there are optimal degrees of

interactivity that differ depending upon the objectives and the nature of the specific multimedia production. Increasing interactivity beyond a certain point can be extremely costly, and the benefits may not justify the increased expense. [...] The ultimate form of interactivity in communication is achieved in face-to-face, human communication. [...] Studying interpersonal interaction and applying it to human–computer interfaces in multimedia is important. Developers should always attend to the basic aspects of interpersonal communication in the creation of multimedia productions.

(Tannenbaum, 1998, pp. 301–302)

What kind of interactivity should be studied in order to gain knowledge of human–computer interaction is, however, a matter of debate. The Wizard-of-Oz study by Dahlbäck *et al.* (1993), rejects the belief that human–human dialogue studies are enough to understand how users will behave when interacting with an automaton, even if the users are allowed to use natural language. This ought to have a great impact on the system development process, especially if the system in question is meant to support education and training. How should the didactic requirements for educational software be specified? Should an educational program try to mimic the interplay between teacher and student? Demonstrations and testability of concepts would surely provide system developers with an empirical foundation in each individual case. We believe that Ozlab is easy to use for such purposes.

Because the essential part of a multimedia system is its presentation parts, its extrovert part, it is not possible to adhere to the developer paradigm expressed in many traditional methods. For instance, dialogue design is in SSADM (Structured Systems Analysis and Design Method, version 4; Malcolm, 1994) described as '...an activity that takes place towards the latter part of the analysis and design cycle,' even though it is recognized that preparation begins very early. During the requirements-specification stage of SSADM, dialogues are identified and described in the form of menu and command structures. The actual design of the dialogues is done in the last stage, physical design (Malcolm, 1994). Ozlab permits simple sketches, including interactivity, to be shown and demonstrated, and users/customers are able take an active part in making layouts and running the interaction drives (i.e. by acting as wizards themselves).

Finally, *customers' mental models* of the system/product will vary, or, often enough, will cover only part of the system/product. Possibly, the very fact that the system will interact with the prospective users will make it hard to pinpoint exactly what the product will appear as. Neophytou *et al.* (1995) remarks that the structured analysis modelling tools are not sufficient for modelling multimedia information and that there is a need for modelling tools for the development of multimedia systems. To this should be added the different methods for giving the users/customers an idea of possible solutions (Preece *et al.,* 2002, p. 169). Web site designers often work with outlines in the form of paper sketches (Klee, 2000; Cato, 2001). Here, we draw attention to the suitability of Wizard-of-Oz experimenting for developing graphical and multimedial user interfaces. Displaying interactivity concretely for the user necessarily involves a time-dependent presentation, which is something that a computer-based presentation tool supports. This leads to the next section.

5.4.3 Static models are difficult to use for dynamic content

Diagrammatic models, as used in systems development, are good for describing structure, including data flow, but not for showing time-dependent content. Specifying such content in a static medium is not enough for customers to realize the implications of the written or drawn requirements. How sound and picture go together or the altering phases of an interaction should preferably be experienced before decided upon. Dynamic modelling is used in multimedia *products* in the form of animations. It is reasonable then to propose the use of dynamic modelling to strengthen the dialogue between customer and developer.

Ozlab is one such attempt. It builds on a kind of manual simulation of the interactivity of a multimedia product. The purpose is to find means of expressing and formulating ideas and experiences of interactivity. In a Wizard-of-Oz experiment, the user is confronted with a layout and functionality which appears complete, but the functionality is in fact made possible by a hidden test leader as described in Section 5.3. This makes it possible to test different system responses without implementing them first.

On the other hand, it is difficult to include film sequences or musical pieces in the preliminary tests. Testing—together with the customer/user—different mixes of video clips with background music demands that these things be at hand already during the conceptual work. That is tantamount to collecting the material needed for the final production before it is decided on what the product will include, as we will see in the next section.

5.4.4   The understanding deepens as the work proceeds

Even if we use a demonstration technology such as Ozlab, which captures most of the layout and the function as far as the user is concerned, it is, for various reasons, not always possible to have all the requirements at hand when the provider writes the offer to the prospective customer. It is, furthermore, not only the customer who may later call for modifications of the system requirements, but also the system developers. Both parties must have an understanding of the other party's need to modify the initial specification. At the same time, unsettled specifications are a source of heavy delays and rocketing costs (as noted by many, among others the Standish Group, 1994). However, what the survey shows is that the developers and customers of multimedia systems have to include modifications of the initial specification in the system development. The question is then how to organize the process so that late changes can be included without exceeding the budget. Nevertheless, one also has to tackle the 'opposite' question: is it possible to make a more thorough requirement analysis before the implementation starts? This question will be dealt with in the following section.

5.4.5   Customers have vague notions and requirements

One of the major challenges for both customer and supplier is that customers often have only vague ideas about what they want. The path from a vague idea to a good equivalent of a requirements specification may thus be long, even when the ambition to produce one exists. One alternative to a clear requirements specification is more informal agreements about the content, which are formulated in stages. This is common in the multimedia industry today.

> They help you to formulate them, to get them down on paper. For they can generally answer yes or no. Then it's just a matter of getting them down on paper and helping to formulate them. For it is the case that when you have worked at this a while, although we don't have that amount of experience, you find the same

things keep coming up. So you can ask the right questions and they will answer yes or no and then you more or less know.

(R1)

You have to put together a picture of what you're going to do, partly to understand what the customer wants and also to make sure that we have understood correctly. So it's very important.

(R6)

Demonstration by developers and by customers/users of possible solutions and exact design details, including interactivity, is possible in Ozlab experiments. Thus we sometimes refer to Ozlab as a 'user interface articulator'. Articulating one's notions will make them firmer. Furthermore, and as already pointed out above, it is relatively cheap to make mock-ups in a graphical Wizard-of-Oz system. However, there is one advantage in using simple paper sketches instead of working prototypes in the early design phase. People will make more comments on function and structure than on graphical details when the system does not look complete (Olsson *et al.*, 2001; cf. Cato, 2001). In admitting this, it is worth stressing that in order to give the customer a feeling of the interactivity before the contract is signed, computer-based prototypes are probably to be preferred, especially if this interactivity is based on graphics. (Sometimes they are simply vital, such as when one is testing on young children or on adults with severe learning disabilities.)

What the slash compound 'customers/users' above indicates is that the success of an implemented system depends on more than a precise requirements specification from the customer. Neophytou *et al.* (1995) claim that the hands-on users are either completely left out of the development process, or at best, are involved as a sample group during the test phase. Obviously, a customer with a vague notion of either future users or of the need to include these groups in the definitional work, will not be able to identify requirements a properly.

5.4.6 Difficult to give requirements in explicit and measurable terms

As stated in the opening paragraph of this chapter, one of the arguments for why requirements ought to be explicitly formulated is that they should be *verifiable* (cf. Davis, 1993). A concept closely related to this is that of *measurable* requirements. The benefits of explicit and measurable requirements can be briefly stated as:

- An explicit requirement can be communicated and discussed.

- Measurable requirements are necessary when testing if the requirement is met in the final product.

Multimedia products contain many properties that are difficult to express in explicit terms. Following on this are problems of measurability. The following quote illustrates the difficulty of measuring such properties:

> We want to show similar productions, to show their scope. It's difficult to show the scope of a production. Should you measure how long an interactive education programme takes?
>
> (R2)

To continue, even if a requirement is formulated explicitly it can be hard to measure in objective terms. This complicates its verification later on when the system is delivered. A multimedia presentation of a bank should perhaps give you a feeling of confidence. This is an explicit requirement which is not easy to measure. How can it be verified that such a requirement has been met in the final product? The method of verification has to be decided with regard to the type of requirement. In the bank example, a possible solution would be to use a test panel to measure the degree of confidence in some way. With Ozlab one needs not have the multimedia presentation implemented in order to start the testing. This makes it easy to discuss and develop the verification model.

Requirements pertaining to the performance (in a quantitative sense) of the users of a future system are easy to verify, since they are expressed in numbers and are not based on the opinions of a test panel. Wizard-of-Oz set-ups admit for tests as if the software already existed and thereby allow to some extent for the measurement of productivity or other effects of actual use.

A question closely related to the demand for explicitness is that of how to express what is not wanted. When an acceptable test set-up has been reached, it is in itself a specification for the user interface including its functionality. To show the software developers what is intended, enough material may be taken from automatic logging functions in the

Oz system. Useful material may also be found in video recordings made during test runs. An early suggestion for combining the Wizard-of-Oz technique with 'video prototyping' is found in Mackay (1988). The usefulness of video recording in ordinary usability tests has been recognized by usability experts: 'The reaction of designers to seeing videotapes of users failing with their system is sometimes powerful and may be highly motivating' (Shneiderman, 1998, p. 131). Thus, video shots from unsatisfactory set-ups can probably be informative in showing what is not wanted. In addition, shots from the control room could show why a particular thing was done in a particular way rather than in the way the orderer says he/she wants it done.

Quality of requirements specification as described by Bubenko and Kirikova (1999) and others, provides an important concept for this discussion. However, we will not expound on this concept here. Instead we will develop the discussion as to the suitability of a multimedia communicational mode for requirements specifications.

5.5 WRITING REQUIREMENTS VS VISUALIZING REQUIREMENTS

The Wizard-of-Oz technique would, as argued above, make it easier to obtain something concrete out of loose ideas. It would be possible to try out initial sketches, even of interactivity, and that should lead to refinements and resolution of contradictory demands. As noted in Section 5.4.6, when an acceptable set-up (including a scheme for its interactivity) has been reached, it is in itself a specification for a product. Virzi *et al.* (1996) point out that prototypes, especially high-fidelity ones, are useful 'in helping the development team to read and understand a detailed specification'.

This citation is interesting because not only does it point to the usefulness of prototypes in mediating requirements, but also it reveals that professional developers may find it hard to 'read and understand a detailed specification'. To compare, Molich (2002) argues, in a handbook for web designers, that a prototype can often replace the requirements specification when developing a web site. To what extent a prototype can, in fact, replace the requirements specification is an intriguing question. In part it depends on whose responsibility it is to set data communication requirements and other non-surface features in an appropriate way. For a customer with no computer expertise, the system provider will also have to provide this expertise or the customer will have to rely on a third-party expert. The latter solution has the disadvantage of splitting the responsibility for the final system onto two parties. For the purpose of the present chapter, it should be noted that the sub-surface part of the system is not part of the defining criteria for 'multimedia', because nothing is 'multi mediated' to a living human being unless it surfaces in the user interface. Of course, it has to be admitted that bandwidths, data-base design, and platform compatibility will affect the user inter-face functionality (for data bases, see Axelsson, Chapter 7 in this book). However, this fact only makes it more interesting to investigate to what extent a prototype could function as the specification for a future system.

This question will get different answers depending on what kind of multimedia system one is trying to specify. If a lot of material is to be stored in the final system (e.g. sound effects or videoclips), it will be hard to prototype every single piece of the future system, because that would presuppose the existence of the material beforehand. In this case, it would be better to have a contract stating how negotiation of content should be done during the development. If the focus is on inter-activity, as in educational software and games, then less new material is needed. Still, however, there is the issue of how much testing should be done before programming takes place. Action games would be hard to simulate without an extensive, programmed prototype.

The answer to the question thus depends on how much it is possible to demonstrate, and also, as noted in Section 5.4.6, how well unwanted fea-tures could be identified. There are no self-evident methods here, even if demonstrations that include interactive features could take us a long way when it comes to defining multimedia products. A demonstrator provides the right medium for specifying a prospective product, which will depend on sound, image, and interactivity. One could compare the 'Contextual video for learning' project (Björgvinsson and Hillgren, 2001, 2002), run by the Interactive Institute in Malmö, Sweden, at the University Hospital of the same town. In order to remind nurses at the intensive care unit of how to use various machines and other utensils, the nurses make short video films which could be accessed via handheld devices. In this way they do not have to read a manual that might be rather cumbersome in a stressed working situation. A manual—even a short instruction tag—may also be cumbersome to write, since many practical things are hard to describe in words while they may be very easy to demonstrate.

This example clearly indicates the utility of choosing the medium in accordance with what should be mediated. The machines and other utensils in the intensive care unit are physical objects with some spatial properties. Spatially arranged physical objects imply pictures as the medium, especially photographs, while how-to-do routines imply sequences, possibly temporally based. Hence, film is suggested. To this is added the tradition among nurses to show each other how various things should be done. This suggests voice recording and thus makes the case for film even stronger.

However, the example is not quite parallel to requirements engineering work. A somewhat closer parallel is provided by The *Visual Assistant*, a program developed to enable students to visualise scenes in a theatre play. 'It provides an imaginary space into which two-dimensional images can be placed and manipulated in three dimensions [. . .] in order to quickly create visualisations relevant to performance.' (Beardon and Enright, 1999). The system is simple to operate; it works by drag-and-drop actions. In computer terms, the system could be used to produce some kind of prototype for the design of scenes in a play. Exactly what purposes the *Visual Assistant* could serve is, however, not quite settled yet:

> At one level it can be seen as a more general visualisation tool for theatre. It is a 3D sketchpad/scrapbook: a way of working with images to help us visually understand a play; or a way of communicating simple ideas about a performance in visual form. That may be as far as one wants to take it. At another level, this can be seen as a preliminary step in a longer process to designing a real scene. It is a way of working out ideas without committing oneself to the time involved in building significant models.
>
> (Beardon and Enright, 1999, p. 163)

In this context the inventors refer to *virtual_Stages* (Dyer, 1999), a 3D graphics stage visualizer, but it suffices here to conclude that the use of a simple demonstrator assistant is useful but needs human interpretation before the sketches become an implementation on either *virtual_Stages* or on a real stage.

Nevertheless, as a medium for discussion between developers and end-users, a visualizer can be very important. To connect to the three points opening the chapter, a visualizer can serve as part of the foundation for design and construction (this was the second point). The extent to which it could be used to settle the requirements specification and provide verification measurements is not clear yet (points 1 and 3). More data, and more development of visualizers, are needed. However, the question as to what extent a prototype could function as the specification for the future system has a legal aspect also. This issue is worth raising now. The above parallel drawn from the theatre arts cannot easily illustrate this aspect, and the following review of architects at work shows the point better.

5.5.1 Resolving disputes

Architects use both sketches and proper drawings to communicate their ideas to customers and also to develop and test their ideas. Letters and numbers are added to the drawings to clarify the nature of material and to give measurements. However, it is the drawings rather than what is written that best describe the constructions planned by architects. Three-dimensional spatial relations are best rendered with spatial relations, even when the rendering is two-dimensional. The British linguist Peter Medway, who has studied architects' choices of media for communication in different work contexts, says 'My observation of and interviews with practising architects leave no doubt that the graphical mode is the one they feel most at home with. Architects generally dislike writing' (Medway, 1996).

Medway notes that much speech and gesture can accompany drawings. Nevertheless, he notes that the time spent on writing increases among architects in Canada and in North America in general: for legal reasons documentation is needed, and documentation demands a medium where the signs are permanent. 'Drawings are permanent too, of course, but in case of dispute are overridden, at least in Ontario law, by written specifications' (p. 29). He also discusses the expressiveness of the two permanent media and points out that 'there are realities that architects need to refer to that drawings cannot, or cannot quickly and economically, represent: for instance, the activities of the users of buildings (say a sequence of actions to prepare omelettes in a kitchen) [. . .] and construction processes [. . .]' (p. 36).

Perhaps a change will come now that computers are more often used for visualization through 3D animation. Showing how a certain part of the building should be built or showing how ergonomically faithful representations of human beings move in a kitchen—this is entirely possible, and such techniques have, since the end of the last century, begun to be used by architects to give different stakeholders a feeling for various alternatives (see, for example, the discussion in Tasli and Sagun, 2002). Still, legal regulations definitively emphasize another medium of communication:

> The quantity of written documentation has doubled in the professional lifetimes of Canadian architects still only in their forties. Record-keeping in Europe is not so demanding, but may become so as a result of changing European Union regulations.
>
> (Medway, 1996, p. 29)

It is too early to say whether or not the possibility of visualizing competing solutions will change the way the relationship between image and writing is weighed in legal disputes. For multimedia systems development based on Wizard-of-Oz set-ups, it is relevant here to consider again measurable requirements. Such requirements may not be visualized themselves. Even if it is the visualization that makes possible an early setting of minimum requirements for the degrees of user acceptance and performance, these requirements might take a verbal form. It is true that an Oz set-up does not merely show on a superficial level what the ordered system would look like, but actually how it functions. It is also true that this functionality is demonstrated in rather explicit ways. However, such a specification does not automatically generate measurable requirements. Customer and provider have to agree on what to measure if the final product is meant to be compared with the Oz files. Hopefully, the early experience of the proposed system has made the customer aware of two things: one is what he/she regards as the essential features of the proposed system, the other is how the existence of these features in a future system should be verified.

5.6 REQUIREMENTS SPECIFICATION BY CONTENT PROFESSIONALS

Claiming that Ozlab is an easy-to-use, combined usability testlab and workstation for explorative interaction with users, entails a further claim—or at least a goal—that such a laboratory can be used by lay-men to define the design of the future system. Simple multimedia tools for school use (e.g. *HyperStudio*, *Creator*, or Swedish *MultiMediaLab*) show that non-programmers can take on the role of designers-and-implementers, at least when the systems built are fairly simple. More-over, non-programmers are professionals in their own field of work and could have some advantages over the professional designers. For inter-active teaching aids, it is not a good idea to leave it for the professional designers to define the interactivity of the product. Not only content but also how content is mediated is at the very core of the matter and the professional designer is not an expert on the content. Moreover, he/she is probably not even fully aware of the end-users' (e.g. children's) cognitive abilities. Computer support for people with special needs has indeed been developed in close collaboration with therapists, teachers, and researchers—often the programmers have also belonged to these groups.

However, not every therapist or educationalist is a programmer and there are many ideas that will not reach even the specification stage. We will conclude our chapter by sketching a procedure for the lay require-ments specification.

As noted, there is the potential of Ozlab to be used by multimedia laymen. The Oz mock-up is not a functioning prototype. To really show the software developers what is intended, ample material must be taken from video recordings made during test-runs and logging-functions in the Oz system. Such a product specification would lessen the demands on the inexperienced designer to know how to express the requirements. To formulate measurable requirements according to trade standards, naive designers may need the help of the contractor or some other expert. This might sound a straight forward procedure, but the exact nature of the expertise might need consideration. Programmers are not always helpful, as five professional developers claimed in another interview survey of multimedia developers:

> They even claim that it is common that programmers put restrictions on the con-tent producers. To begin with, these restrictions may be necessary because of technical considerations, but in some cases these restrictions continue even if the technical problems have been solved. In order to avoid this, [the company] has changed approach from a technically oriented approach to a more content-oriented approach. Now, the content producers create and design solutions without

reflecting on the technical solution. The technicians then do their best to implement the content. Since [the company] changed the approach, the company has made the fastest technical progress ever.

(Jonasson, 2002, p. 614)

Such a story clearly speaks in favour of a more implementer-independent design phase (cf. Löwgren, 1995). Nevertheless, a feasible procedure for educationalists and others with access to a working Ozlab may be the following, which includes an expert for pin-pointing requirements and providing suggestions for design alternatives:

(1) Ideas stemming from their own practical work result in an initial sketch (= Oz sketch).

(2) Oz testing of their own design.

(3) Multimedia expert (perhaps specializing in the field in question, such as programs for special needs) is hired for evaluation and suggestions concerning requirements specification as well as design alternatives.

(4) Oz test of new design.

(5) Pack (all forms of) requirements specification for cost estimation or quotation.

Stage 3 is associated with some costs but, on the other hand, the consultant does not have to start working from scratch. In all, this procedure leads to a cost-efficient and more professional specification. The consultant in stage 3 should not only help with formulation of different written requirements specifications, but also in the selection of ample video and screen-dump samples. Again, video shots from unsatisfactory set-ups can probably be informative as well as shots from the control room.

In stage 5, the word 'pack' is intended to emphasize the role of the video tapes and computer log files, in addition to the mandatory written document, even if the latter may contain drawings of the graphic layout. Stage 5 may lead to a contract, which should also then include this non-written documentation.

As yet, we are far from measuring the quality of such specifications. Rather, an initial pilot study which ran during autumn 2001 focused on the feasibility of Ozlab tests performed by non-programmers. Three educationalists partook and in some trials also their clients (in this case young children with Down's syndrome). However, the interesting observations on how they matured in their role as wizards (i.e. as stand-ins for a real system) are outside the scope of the present chapter and has been reported elsewhere (Pettersson and Sone, 2001; Pettersson 2002). Likewise, Ozlab has been used in human–computer interaction courses and in students' projects, as reported in Pettersson (2003).

5.7   CONCLUDING REMARKS

What this chapter addresses is the need to find ways of articulating interactivity, that is, the need to find means for expressing ideas and experience of interactivity. The samples from the interview survey shows that orderers of multimedia systems most often lack the basic knowledge needed to formulate the requirements of the system they want to order. The developers interviewed met this problem by developing the systems with more or less frequent contact with the customers.

This chapter presented a graphically based Wizard-of-Oz system, Ozlab, as a tool for discussion with the customer, but it also stressed the need to include the final users. It was recognized that simple paper mock-ups permit early evaluation of design concepts and has been recommended by usability experts. However, it was pointed out that there are circumstances when meaningful tests cannot be performed unless more realistic interactive products are used. The versatility of an interaction articulator was therefore suggested in relation to common problems of multimedia development.

The pilot test briefly mentioned at the end of this chapter included children having a varying degree of abstract thinking. With Ozlab, programming is avoided while the mock-up looks and behaves like a real computer program. Therefore, testing various proposals for interactive products are meaningful even with this target group in mind. Furthermore, the presented Ozlab system can also be used by non-professional designers to experiment with, and to some extent to evaluate, their own ideas and discover what kind of interactivity suits a certain purpose and target group.

The challenging question as to whether or not multimedia-requirements work and specification should be done multimedially or not will, however, depend very much on the nature of the multimedia product developed. A system such as that above, which allows test-running mock-ups of interactive multimedia products, will not allow very speedy interactions like those found in action games. Similarly, if a product is to include very many multimedia samples, then there is hardly any point in collecting them all before production starts. Only some typical examples need be given in advance as a means of specifying the product. For these reasons, the actual practice of multimedia developers, as revealed in the interview survey, will sometimes be the only possible procedure. Checkpoints have to be included in the development, which starts after the contract has been signed. The problem with this is that disputes may arise on the way and the parties—provider and customer—will be left at each other's mercy. This has not been much highlighted when it comes to multimedia projects, but for ordinary systems development, examples of systems developments surpassing time and budget limits have been frequent (e.g. the CHAOS report from the Standish Group, 1994). From the parallel of North American architects' dependency on written documents rather than drawings, one must realize that an early prototype covering all aspects of the surface properties of a multimedia system may not settle disputes. The role of visualizing tools like Ozlab in contract writing may, however, be that of setting measurable criteria for the final system to meet.

5.8 REFERENCES

Beardon, C., and T. Enright (1999) 'Computers and Improvisation: Using the Visual Assistant for Teaching', *Digital Creativity*, **10**(3), 153–166.

Björgvinsson, E.B. and P.-A. Hillgren (2001) *Kontextualiserad video för kontinuerligt lärande inom vården*. Presented at STIMDI-01, Solna, 22–23 October 2001. Available at <http://www.stimdi.se/arrangemang/konf/stimdi01/artiklar/kontext2.pdf>

Björgvinsson, E.B. and P.-A. Hillgren (2002) 'Readymade Design at an Intensive Care Unit', in *Proceedings from Participatory Design Conference*, June 2002, Malmö, Sweden.

Blum, B.I. (1996) *Beyond Programming—To a New Era of Design,* Oxford University Press, New York.

Bubenko Jr, J.A. and M. Kirikova (1999) 'Improving the Quality of Requirements Specifications by Enterprise Modelling', in *Perspectives on Business Modelling—Understanding and Changing Organisations,* A.G. Nilsson, C. Tolis and C. Nellborn (Eds) Springer-Verlag, Berlin, pp. 243–268.

Caelen, J. (1999) 'MultiCom, a platform for the design and the evaluation of interactive systems', *Invited conference at Al Shaam'99 congress.* Damascus, April 1999. Proceedings, pp. 1–11. Available at <http://www-geod.imag.fr/jcaelen/>

Caelen, J. and E. Millien (2002) 'MultiCom, a Platform for the Design and the Evaluation of Interactive Systems: Application to Residential Gateways and Home Services', *Les Cahiers du Numérique,* **3**(4), 149–171.

Cato, J. (2001) *User-Centred Web Design.* Addison-Wesley, Harlow, UK.

Coutaz, J., L. Nigay, and D. Salber (1995) 'Multimodality from the User and System Perspectives', *ERCIM'95 workshop on Multimedia Multimodal User Interfaces,* 1995 Crete.

Dahlbäck, N., A. Jönsson and L. Ahrenberg (1993) 'Wizard of Oz Studies—Why and How', *Knowledge-Based Systems,* **6**(4), 258–266.

Davis, A. M. (1993) *Software Requirements—Objects, Functions and States,* Prentice Hall, New York.

Dyer, C. (1999) 'virtual_Stage: An Interactive Model of Performance Spaces for Creative Teams, Technicians and Students', *Digital Creativity,* **10**(3), 143–152.

Isaacs, E. A., T. Morris and T. K. Rodriguez (1995) 'Lessons Learned from a Successful Collaboration Between Software Engineers and Human Interface Engineers', In O. Taylor and O. Coutaz (Eds) *Software Engineering and Human-Computer Interaction,* Proceedings of the ICSE '94 Workshop on SE-HCI: Joint Research Issues. Springer-Verlag, Berlin, pp. 223–240.

Jonasson, I. (2002) 'Future Trends in Developing Multimedia Information Systems—Competencies and Methodologies', *PROMOTE IT 2002*, 22–24 April, Skövde, J. Bubenko Jr. and B. Wangler (Eds), pp. 609–620.

Klee, M. (2000) 'Five paper prototyping tips', *UIE Newsletter Eye for Design*, March/April. Also at <http://world.std.com/~uieweb/paperproto.htm>

Löwgren, J. (1995) 'Applying Design Methodology to Software Development', in *Symposium on Designing Interactive Systems*, DIS'95 Proceedings, ACM Press, New York, pp. 87–95.

Mackay, W.E. (1988) 'Video: Data for Studying Human–Computer Interaction', Panel discussion at *CHI'88 Human Factors in Computing Systems*, May 15–19, 1988, Association for Computing Machinery, pp. 133–137.

Malcolm, E. (1994) *SSADM Version 4: A User's Guide*, Second Edition, McGraw-Hill, London.

Medway, P. (1996) 'Writing, Speaking, Drawing: The Distribution of Meaning in Architects' Communication', in Sharples, M. and T. van der Geest (Eds), *The New Writing Environment: Writers at Work in a World of Technology*, Springer, London, pp. 25–42.

Molich, R. (2002) *Webbdesign med fokus på användbarhet* (trans. from Danish Brugervenligt webdesign, 2000), Studentlitteratur, Lund.

Molin, L. (2002) *Multimediautveckling—konst eller teknik? Reflektioner kring produkt, process, kravarbete och kunskapsutbyte.* [Multimedia development—Art or Technology? Reflections on Product, Process, Requirements Specification and Learning]. Licentiate thesis. Department of Information Systems, Karlstad University, Sweden.

Neophytou, A., G.A. Papadopoulos and C.N. Schizas (1995) 'Multimedia Systems Development: The Use of Existing Modeling Tools', in *Proceedings of the Third European Conference on Information Systems*, ECIS'95, Athens, Greece, pp. 721–733.

Olsson, S., F. Denizhan and A. Lantz (2001) *Prototyping* [in Swedish] TRITA-NA-D0105, CID. Kungliga Tekniska Högskolan, Stockholm. <http://cid.nada.kth.se/sv/publikationer/alla>

Pettersson, J.S. (2002) 'Visualising Interactive Graphics Design for Testing with Users', *Digital Creativity*, **13**(3), 144–156.

Pettersson, J.S. (2003) 'Ozlab—A System Overview with an Account of Two Years of Experience', In J.S. Pettersson (Ed.), *HumanIT 2003, Karlstad University Studies*, Vol. 26, Centre for HumanIT, Karlstad, pp. 159–185.

Pettersson, J.S. and T. Sone (2001) Ozlab—ett användarvänligt användarlabb i Karlstad. Poster presentation at STIMDI-01, Solna, 22–23 October 2001. Available at <www.cs.kau.se/~jsp/ozlab>

Preece, J., Y. Rogers and H. Sharp (2002) *Interaction Design*. John Wiley & Sons Ltd, Chichester, UK.

Promise (2001) 'Producers of Interactive Media in Sweden' <http://www.promise.se/> (inspected 26-11.2001).

Shneiderman, B. (1998) *Designing the User Interface Strategies for Effective Human—Computer Interaction*, Addison-Wesley, UK.

Standish Group (1994), 'CHAOS report', <www.standishgroup.com/sample_research/ chaos_1994_1.php>

Tannenbaum, R.S. (1998) *Theoretical Foundations of Multimedia*, Computer Science Press, New York, USA.

Tasli, S. and A. Sagun (2002) 'Proposals for Creative Uses of Computer Graphics in Architectural Design', *Digital Creativity*, **13**(3), 182–185.

Virzi, R.A., J.L. Sokolov and D. Karis (1996) 'Usability Problem Identification using both Low- and High-fidelity Prototypes', *Conference Proceedings on Human Factors in Computing Systems* CHI'96, April 13–18, Vancouver, Canada, pp. 236–243.

6

Evaluating Interactive Multimedia in an Inexpensive and Timesaving Way–Illustrated by Business Case Studies

Louise Ulfhake
Information Systems, Karlstad University

6.1 BACKGROUND

Both researchers and practitioners are convinced that evaluation is a very essential part in the development and maintenance of information systems and their products. Evaluation is important in order to assure quality during the evolution of interactive multimedia as well as in reviewing

Perspectives on Multimedia R. Burnett, Anna Brunstrom and Anders G. Nilsson
© 2004 John Wiley & Sons, Ltd ISBN: 0-470-86863-5

of the final product. England and Finney (1999) tried to define quality in multimedia, but they ended up with a broad, generic sentence: 'Design quality for media projects = Content and treatment agreement'. They found that there are objective parameters for some of the technical aspects but there are many subjective aspects for design quality. Despite that, it is necessary to evaluate multimedia production in some way. Large companies often possess a test and evaluation department but small and medium-sized companies cannot afford that. A systems development method for multimedia applications ought to involve an evaluation stage, according to Ulfhake (2001). However, in any case, the evaluation should be done preferably in a very quickly way and, above all, free of charge or, at least, inexpensive. This contribution proposes an inexpensive and timesaving evaluation method that suits even small and medium-sized enterprises that are developing educational multimedia applications.

In the years 1998 to 2000, the European program ESPRIT IV financed a project named BUSINES-LINC (*Bus*iness *I*nnovation *N*etworks— *L*earning with *I*nteractive *C*ases). One of the objectives of the project was to develop multimedia case studies. Seibt (2000) writes:

> The rationale of project BUSINES-LINC was to support Business Innovation Processes within European Companies through the provision of *interactive and inter-linked innovation case studies.*

According to Bielli and Basaglia (2000) a case study is 'the description of a real situation *cut* or *interpreted* in relation to specific teaching purposes'.

Stockholm School of Economics in Sweden was one of the BUSINES-LINC project's members.[†] In the autumn of 1998 they finished their first out of three productions. About a year later the other two multimedia case studies were finished, having been based on the first production's experience and evaluation. I was involved to do the evaluation of all three multimedia productions.

[†] Project members were University of Cologne (Project Coordinator), Copenhagen Business School, Erasmus University in Rotterdam, Norwegian School of Economics and Business Administration in Bergen, Bocconi University in Milano, and Stockholm School of Economics.

6.2 THE MULTIMEDIA CASE STUDIES

A primary goal of all cases was to highlight and assist the transfer of practical innovative and technological solutions for different industries. The cases therefore describe successful innovation in these industries. They are presented on CD-ROMs and in formats intended by designers to allow undergraduates, MBA students and executive program participants to obtain useful insights (ECCHO, 2000). The cases are intended to be used within a pre-defined educational framework, a process with a mix of lectures, assignments and workshops. They are therefore intentionally rich, i.e. they are able to be used for more than one theme, for instance, different aspects of the presented business solution or the change process.

Busines-Linc's first Swedish production was completed at the end of 1998 and that one is *Sandvik Saws and Tools—the New Distribution Concept*. It explores the successful transformation in distribution achieved by Sandvik Saws and Tools, a Swedish manufacturer of professional hand-tools and saws. The designers used the adventure game analogy with an overall game plan. The game plan was set up geographically, which means giving an overview of the different locations involved in new distribution process. In this way, it is possible to go through the case by talking to some people, getting some papers and brochures, and getting a tour of some part of the company.

The other two Swedish productions are *WM-data Product Supply and WebDirect—'800 million reasons to use the Internet'*, and *Shaping a Virtual Bookstore—the story about Bokus.com*. They were completed at the end of 1999. *WM-data Product Supply and WebDirect* describes the ideas and the organization behind business-to-business using e-commerce. *Shaping a Virtual Bookstore* presents the way in which the company Bokus was created in just a few months. The designers in these cases also used an overall game plan, but it was set up differently depending on the purpose of the cases. In the case of *WM-data Product Supply and WebDirect*, the game plan was divided into four parts: the setting, the business, the journey, and the key aspects. In the case of Bokus, the game plan consisted of all persons and business partners who were involved in the creation of the virtual bookstore.

When I received the first production[†], together with a six-page document intended for the BUSINES-LINC project, the production was regarded as completed, but the designers of the project wanted it to be evaluated as if it was still active, because they had two more productions to develop. So, when the project received the feedback from the Sandvik case, they integrated the different evaluation criteria into their ongoing development of the next two cases. As the analysis of the business, the interviews and other data collection and the design/construction of the case were carried out simultaneously, it was too late to apply these evaluation criteria at the end of the development process. This emphasizes the need for a simple, but sufficient set of rules because everyone in the project had to be aware of these requirements in order to work productively. As the Sandvik case was developed as a prototype for training the project group in the development of multimedia material for the type of teaching/learning mentioned above, the evaluation feedback was important and useful for the coming two cases but the Sandvik prototype was not updated. Therefore, the project anticipated that the evaluation results should influence their future productions. At that time they had no proposals about any preferred evaluation method.

6.3   EVALUATION METHODS

In the literatures (See Boyle, 1997; Dix *et al.*, 1998; Fowler, 1998; Jerkedal, 1999; Löwgren, 1993; Newman and Lamming, 1995; Preece *et al.*, 1994; Shneiderman, 1998) there are several evaluation methods to choose between. Some of them are intended for the development stages, i.e. formative evaluation, others are more suitable in later stages, i.e. summative evaluation. Formative evaluation has the purpose of obtaining feedback for the next iteration for on-going projects, while summative evaluation is used to test and judge the whole design of a completed system or program (See e.g. Boyle, 1997; Jerkedal, 1999; Löwgren, 1993; Newman and Lamming, 1995 and Preece *et al.*, 1994). Both types of evaluation feed back to influence design practice (Boyle, 1997).

[†] The first production was evaluated in the beginning of February 1999 (Ulfhake, 1999)

Figure 6.1 User involvement in evaluation

Many authors (See e.g. Boyle, 1997; Dix *et al.*, 1998; Fowler, 1998; Jerkedal, 1999; Lif, 1998; Löwgren, 1993; Newman and Lamming, 1995; Preece *et al.*, 1994 and Shneiderman, 1998) try to categorize the evaluation methods in some way, mostly depending on when it takes place (in the design phase or in the implementation phase) or where it takes place (in a laboratory or in the field). Lif (1998) separated methods for evaluation of user interfaces into two groups depending on the user involvement: *usability testing methods* and *usability inspection methods*. He states that usability testing methods involve users while usability inspection methods don't.

It is of great importance to know whether or not you need user involvement and to what degree before choosing an evaluation method. Some evaluation methods cannot be used without cooperation with end-users, while lone experts may use others. In Figure 6.1, some evaluation methods are exemplified, where the user's involvement has different significance.

The expert reviews have the lowest degree of user involvement. In these methods you don't need any users at all to fulfill the evaluation. Expert reviews belong to group usability inspection methods and can be conducted rapidly and at short notice (Shneiderman, 1998). Heuristic evaluation is an example of a usability inspection method and in this case a small group of usability experts evaluates a design using a specified set of design rules. Research results indicate that three to five experts are enough to find most of the usability problems in a design (Nielsen and Mack, 1994). If the experts also know a lot about the application domain, even fewer of them might be enough (Löwgren, 1993).

Usability testing is an experimental method involving end-users (Nielsen and Mack, 1994). The tests are conducted in laboratories or as field studies, and the evaluation is determined by measuring some attribute of the evaluator's hypothesis (Shneiderman, 1998; Preece *et al.*, 1994). Experimental methods (formal methods) have not been widely used in the evaluation of multimedia systems (Boyle, 1997). Sometimes the term 'user testing' is used synonymously with usability testing as by Preece *et al.* (1994) or Lopuck (1996). Lopuck (1996) states that user testing is extremely important and that it should occur during the design and development process of multimedia. Preece *et al.* (1994) write: 'As a general rule, any kind of user testing is better than none.'

Participative evaluation is a user-participative method and it originates from socio-technical research (Newman and Lamming, 1995). Cooperative evaluation and participative evaluation are two slightly different user-participative methods, where the degree of user involvement and user control is the main difference (Preece *et al.*, 1994). Participative evaluation and participative design share the same philosophy and they are basically one and the same method. Shneiderman (1998) states:

> ... that more user involvement brings more accurate information about tasks, an opportunity for users to influence design decisions, the sense of participation that builds users' ego investment in successful implementation, and the potential for increased user acceptance of the final system.

He points out that participatory design experiences are usually positive (Shneiderman, 1998).

What do you do if you can't involve any end-users in your evaluation work? Well, you can use any inspection method, for example expert reviews (Preece *et al.*, 1994). Experts can differ in experience and skills, e.g. in the application domain, in the user interface design and evaluation, in the instructional or educational design and teaching. Nielsen and Mack (1994) noticed that:

> Not surprisingly, experience with usability guidelines, user testing and the design of user interfaces leads to more effective usability inspection problem reporting. This is compared to inspectors who are software developers but lacking interface or usability expertise, or inspectors who are not experts in either software development or usability.

Due to the lack of time and other resources, it was impossible to involve any end-users in the evaluation of the multimedia case studies mentioned above. The evaluation method I chose, as an expert in software development and usability, therefore, was expert review using heuristics.

6.4   WHAT TO EVALUATE

Your first encounter with an interactive multimedia product is by its structure, in other words, how the whole production is organized. As Kristof and Satran (1995) write:

> As information designer, you are a gatekeeper. Even though users make their own choices, it's up to you what choices they have—what they see first, where they can go, and what they don't see at all.

Lopuck (1996) stresses that it is necessary to understand the structural possibilities of multimedia before you can effectively design a multimedia production, and, according to Minken and Stenseth (1998), you can gather the designer's pedagogical approach by looking at the production's structure. Evaluation of the structure, is, therefore, your first consideration when looking at a multimedia production.

Your next consideration includes the interaction between you and the computer. The interaction can be seen as a dialogue between the computer and the user (Dix *et al.*, 1998). The interaction design of the dialogues is very crucial when you develop an interactive production and therefore needs to be evaluated separately, especially if the production is aimed for educational use.

It is not enough to look only at the structure and the interaction of the system. The systems need to be usable for the user. In question is the system's ability to support the user's activities, called usability by Newman and Lamming (1995). In human–computer interaction (HCI), usability is a key concept and every HCI researcher and author uses it, discusses it and measures it.

The developers are also interested in decreasing the cost and increasing the effectiveness, while users want the systems to be useful and support them in doing their tasks (Preece *et al.*, 1994). The productivity of the multimedia production needs also to be considered and evaluated.

In the following parts of the chapter my evaluation method is presented. It is built up by my practical experiences of evaluating educational multimedia applications and what has been found in the literature. The method consists of four consecutive steps and suggests *how* to evaluate an educational multimedia production. The steps are:

(1) Evaluation of the structure, where the organization of the information is looked at.

(2) Evaluation of the interaction, where the dialogue between user and computer is looked at.

(3) Evaluation of the usability, where the usability is looked at from different perspectives.

(4) Evaluation of the productivity, where the production is looked at from a holistic perspective.

6.5   EVALUATION OF THE STRUCTURE

There are many different structures to consider. Lynch and Horton (1997) have aggregated different structures into four basic organization patterns, namely sequences, grids, hierarchies and webs. The chart below (Figure 6.2) compares the four basic organization patterns against the 'linearity' and the 'complexity' of the information (Lynch and Horton, 1997).

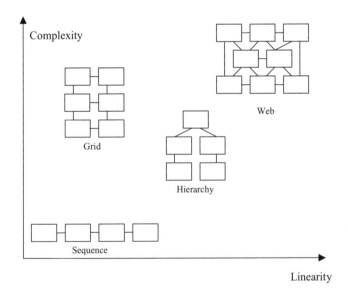

Figure 6.2 Different ways of organizing information according to Lynch and Horton (1997)

According to Lynch and Horton, the simplest way to organize information is to put it into a *sequence*, where the information is presented as a linear narrative. They also state that simple sequential organization usually only works for smaller sites, because long narrative sequences often become more complex, and therefore need more structure to remain understandable. As Minken and Stenseth (1998) point out, the sequential structure of the product implies imperturbable pedagogy.

Lynch and Horton (1997) state that *grids* can be difficult to understand if the user does not recognize the interrelationships between categories of information. Grids are probably best for experienced audiences who already have a basic understanding of the topic and its organization.

The *hierarchical* organization implies that a data structure or logical order of themes has been determined (Minken and Stenseth, 1998). So, if the information is complex, hierarchies are one of the best ways to organize it because most users are familiar with hierarchical diagrams, and find the metaphor easy to understand as a navigational aid (Lynch and Horton, 1997). Users, therefore, find it easy to build conceptual models of the site.

Web-like structures are the most complex ways of organization and Lynch and Horton (1997) stress this in the following sentences:

> Web-like organizational structures pose few restrictions on the pattern of information use. The goal is often to mimic associative thought and free flow of ideas, where users follow their interests in a heuristic, idiosyncratic pattern unique to each person who visits the site chunks.
>
> (Lynch and Horton, 1997, site structure p. 2).

They also point out problems with the structure:

> ...but web-like organization structures can just as easily propagate confusion and fuzzy thinking about the interrelationships of your information chunks.
>
> (Lynch and Horton, 1997, site structure p. 2)

According to Minken and Stenseth (1998), this open structure can imply an independent and open-minded pedagogy and they prefer this way of thinking and structuring productions. However, according to Lynch and Horton (1997), webs work best for small sites dominated by lists of links, aimed at highly educated or experienced users looking for further education or enrichment, rather than for a basic understanding of the topic. They (Lynch and Horton, 1997) and Shneiderman (1998) have noticed that most complex web sites share aspects of all four types of information structure, and that graphic overview maps, so-called site maps, are very useful in helping the user to build a 'real' conceptual model over the production.

Minken and Stenseth (1998) use the metaphor *marketplace* when they describe the structure of a production, and this structure is less complex than that of the web, but still open minded. The marketplace gives a visual overview of the production's functions and the user's possibility of actions, because the marketplace consists of an open place and a number of separated stalls. The structure can be made more complex by permitting connections between different stalls, which gives it a web-like look.

The structure of *Sandvik Saws and Tools—the New Distribution Concept* is described by using the metaphor marketplace (see Figure 6.3).

The 'First Site' is the opening screen and from there you can choose to go to credits about the authors, information about the project (Info. Busines-Linc) or to the case information and case contents. From the stalls 'Info. Busines-Linc' and 'Dictionary' you can leave the case and enter other sites on the Internet, without observing that. The case information part consists of:

- *Analytical Toolbox*, where an analytic tool is introduced and a framework for studying on-going or finished business innovation cases is presented.

- *Printing the Case*, which was actually not implemented.

- *Using the Case*, where a brief overview of the production and a recommended route through the case are presented.

- *Introduction*, where the background of the new distribution concept at Sandvik Saws and Tools is presented.

- *Dictionary*, which is a case glossary with links out on the Internet.

The 'Case contents' of the 'Business Case' consists of six stalls, where you can find information presented in for example text, video, table, and illustrated sound. The six stalls are Corporate Headquarters, Project Office, Distribution Center, Product Center, National Sales Units and Customers.

In the production *Sandvik Saws and Tools—the New Distribution Concept* the whole organization is web like, because you can reach all

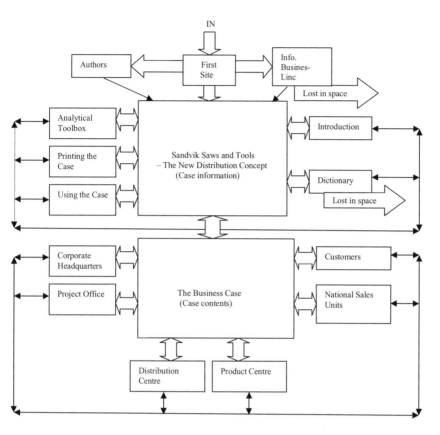

Figure 6.3 Structural description of the case 'Sandvik Saws and Tools', according to the Marketplace metaphor

other stalls from any stall, as indicated by the thick arrows in Figure 6.3. The production consists of three main layers. The thin arrows show that you can go from inside one stall to any other stall at the same level. The structure inside each stall is of both hierarchical and sequential organization, which indicates that the production is very complex. Due to the complexity of the structure and the absence of a site map, it is difficult to build a conceptual model of the production.

The other two productions (*WM-data Product Supply and WebDirect—'800 million reasons to use the Internet'* and *Shaping a Virtual Bookstore—the story about Bokus.com*) also have a web-like structure, but they include site maps and consist of only two main layers. It is therefore easier to build conceptual models of them and to work with them.

Problems with building a conceptual model depend on more than just the structure of the system. The relationship between designer's model and users' model is an important consideration of conceptual models (Preece *et al.*, 1994). Lack of hardware or software functionality might also cause confusion, because it is not possible to bring to an end the conceptual model of the production.

6.6   EVALUATION OF THE INTERACTION

Interaction is the communication between human and system. Interaction can for that reason be seen as a dialogue between the computer and the user (Dix *et al.*, 1998). The dialogue is developed in the interface design and so the interface design needs to be evaluated. According to Newman and Lamming (1995) guidelines as general design principles can be used all the way from identifying design strategies to conducting heuristic evaluations.

There exists a number of general design principles such as Shneiderman's (1998) eight golden rules of interface design or the usability heuristics suggested by Nielsen (1994, 2001). Both of them are described in Table 6.1. As shown in the table, there are six rules and usability heuristics that are equal, and six that are unequal. Newman and Lamming (1995) point out the overlap between Shneiderman's eight golden rules of dialogue design and Nielsen and Molich's (1989) evaluation heuristics. (These are earlier versions of the two general design principles mentioned above.) Many HCI authors stress that it is enough to use one of these two general design principles but I used both together. The reason for that is that more aspects of the interaction are enclosed if you look at several general rules. Consequently, twelve rules/usability heuristics were obtained to evaluate towards.

If heuristic evaluation as performed using these general design principles, not only the interaction but also part of the usability will be evaluated. This step and the next step in my evaluation method overlap, so to be observant in this step means that not so many new problems will be detected in the next step.

Table 6.1 Evaluation grounded on Shneiderman's Eight Golden Rules and Nielsen's Ten Usability Heuristics

Rules or Usability Heuristics Shneiderman (1998)/ Nielsen (2001)	Sandvik Saws and Tools—the New Distribution Concept
Strive for consistency/Consistency and standards	There are some inconsistencies in menu names e.g. 'Case Introduction' calls also 'The New Distribution Concepts', 'The Fragmented Market' calls also 'The Benefits of Sales'. There are also some inconsistency in screen design and the use of colours.
Enable frequent users to use shortcuts/Flexibility and efficiency of use	The system does not allow the users to tailor their own shortcuts.
Offer informative feedback/ Visibility of the status	Yes, the feedback is appropriate and reasonably quick.
Design dialogs to yield closure/. . . .	No, not in the higher levels of the system on account of the web-like structure and no site map; Yes, in all detail parts of the product the dialog yield closure.
Offer error prevention and simple error handling/Help users recognize, diagnose and recover from errors	Not relevant because the system does not permit any changes of information.
Permit easy reversal of actions/User control and freedom	Yes, it is easy to step backwards.
Support internal locus of control/. . . .	Yes, the user makes the system react. It is not the system forcing the user to act, except after looking at the video sequences. The system jumps to a new starting position.
Reduce short-term memory load/ Recognition rather then recall	The system is directly manipulated and therefore built on recognition.

continued

Table 6.1   *(Cont.)*

Rules or Usability Heuristics Shneiderman (1998)/ Nielsen (2001)	Sandvik Saws and Tools—the New Distribution Concept
. . . ./Match between system and the real world	Yes, the words, phrases and concepts are from the user's real world.
. . . ./Error prevention	No problems found.
. . . ./Aesthetic and minimalist design	Fulfilled, because the dialogs do not contain irrelevant or unneeded information.
. . . ./Help and documentation	It is possible to reach the user's guide from anywhere, but you need two clicks to reach further information (submenus to a main menu).

In Table 6.1 I have exemplified the use of the twelve rules while evaluating the production *Sandvik Saws and Tools—the New Distribution Concept* (see Ulfhake, 1999).

The production *Sandvik Saws and Tools—the New Distribution Concept* interacts with the information both on the CD and on the web. The interface style is a combination of menu-based, windowing systems where you directly manipulate the items by point and click. When the user selects an item, the appropriate new page is displayed. Targets at the top of the screen allow the user to go BACK to the previous page, to go FORWARD again after going back, or to perform other more specialized functions that are available at the web. Most multimedia systems have these interface styles (Dix *et al.*, 1998; Newman and Lamming, 1995).

There are few problems detected in the production *Sandvik Saws and Tools—the New Distribution Concept* due to it being a mostly thought-out design and that the system is unchangeable for the user. You can just read, listen and move around in the system.

Even fewer problems have been detected *in WM-data Product Supply and WebDirect—'800 million reasons to use the Internet'* and *Shaping a Virtual Bookstore—the story about Bokus.com* in comparison with the first production *Sandvik Saws and Tools—the New Distribution Concept*. The reason is that most problems detected in the first production were addressed in these productions. These were developed in parallel with each others' and both the structure and interaction are very similar.

6.7 EVALUATION OF THE USABILITY

Usability is a key concept in HCI, and specific usability principles can be used during heuristic evaluation when assessing the acceptability of interfaces (Preece *et al.*, 2002). Usability 'is concerned with making systems easy to learn and easy to use' (Preece *et al.*, 1994). The emphasis in evaluation is to identify as many usability problems as possible, which involves rich qualitative data (Newman and Lamming, 1995). Dix *et al.* (1998) refer to successful interactive systems in the sense that they must be usable. They have structured the presentation of usability principles as three main categories:

- *Learnability*—how easily a new user can to start effective interaction to achieve optimum benefit.

- *Flexibility*—how many ways the user can exchange information with the system.

- *Robustness*—how much support the user can access in order to succeed in accomplishing his or her objectives.

More specific principles are associated with each of the main categories (see Table 6.2). The principles affecting learnability are predictability, synthesizability, familiarity, generalizability, and consistency. The principles affecting flexibility are dialogue initiative, multithreading, task migratability, substitutivity and customizability. The principles affecting robustness are observability, recoverability, responsiveness and task conformance.

Preece *et al.* (2002) state that the goals of usability are effectiveness (effective to use), efficiency (efficient to use), safety (safe to use), utility (have good utility), learnability (easy to learn) and memorability (easy to remember how to use). When evaluating Busines-Linc's Swedish productions from the usability point of view, I used the principles to support usability of Dix *et al.* (1998). It is assumed safe that the goals of usability by Preece *et al.* (2002) may fit as well.

Table 6.2   Summary of principles affecting learnibility, flexibility and robustness (based on Dix *et al.*, 1998)

Categories	Principles	Definition
Learnability	Predictability	Support for the user to determine the effect of future action based on past interaction history.
	Synthesizability	Support for the user to assess the effect of past operations on the current state.
	Familiarity	The extent to which a user's knowledge and experience in other real-world or computer-based domains can be applied when interacting with a new system.
	Generalizability	Support for the user to extend knowledge of specific interaction within and across applications to other similar situations.
	Consistency	Likeness in input–output behaviour arising from similar task objectives.
Flexibility	Dialogue initiative	Allowing the user freedom from artificial constraints on the input dialogue imposed by the system.
	Multithreading	Ability of the system to support user interaction pertaining to more than one task at time.
	Task migratability	The ability to pass control for the execution of a given task so that it becomes either internalized by user or system or shared between them.
	Substitutivity	Allowing equivalent values of input and output to be arbitrarily substituted for each other.
	Customizability	Modifiability of the user interface by the user or the system.

continued

Table 6.2   *(Cont.)*

Categories	Principles	Definition
Robustness	Observability	Ability of the user to evaluate the internal state of the system from its perceivable representation.
	Recoverability	Ability of the user to take corrective action once an error has been recognized.
	Responsiveness	How the user perceives the rate of communication with the system.
	Task conformance	The degree to which the system services support all of the tasks the user wishes to perform and in the way that the user understands them.

When most HCI authors refer to usability, their perspective is that of the developer, but the usability can be looked at from other perspectives as well as that of Preece *et al.* (2002), when they refer to user experience goals. However, these goals are not yet clearly defined and can therefore not be used or analysed.

Allwood (1998), on the other hand, talks about usability from a psychological perspective and he looks at it from the user's point of view. According to Allwood, at least four factors that together determine the usability of a production can be identified:

- *Adjustment*, which means that the program functions, should be designed to adhere optimally to the structure of the task that the user is currently solving.

- *User Friendliness*, a combination of the aspects accessibility, consistency with, and support of, human mental functions, individualizing and means of assistance.

- *User Acceptance*, which refers to the user's motivation to operate the program, and to his or her favorable disposition towards the program.

- *User Competence*, or the user's understanding of, and proficiency in, effective interaction with the computer.

When Allwood (1998) discusses usability and uses the word 'user' he mostly refers to end-users, but a user might as well be a customer (see also Molin and Pettersson, Chapter 5 in this book). I have not separated these types of users even though their answers may be different. The reason is that I think that the end-user's perspective is the most important and relevant here. So, 'user' here means end-user.

When looking at the factors of Allwood (1998) and the principles of Dix *et al.* (1998), some overlapping is evident between, for example, Adjustment and Multithreading, User Friendliness and Substitutivity and Robustness. To get a broader perspective when evaluating the usability, despite of the overlapping, I connected the principles from Dix *et al.* (1998) and the factors from Allwood (1998). I did not want to merge them since they represent different perspectives. In Table 6.3 the usability evaluation of *Sandvik Saws and Tools—the New Distribution Concept* is exemplified.

As shown above, the production is easy to learn (Learning) and easy to use (Robustness and User Friendliness), but is lacking in flexibility (Flexibility and Adjustment; Flexibility and User Friendliness). Concerning the factors User Acceptance and User Competence they are depending on the individual end-users and therefore it is not possible to state any problems, just to give a subjective judgement.

The evaluations of the other two productions (*WM-data Product Supply and WebDirect—'800 million reasons to use the Internet'* and *Shaping a Virtual Bookstore—the story about Bokus.com*) point out a similar lack of flexibility. In these productions, as in *Sandvik Saws and Tools— the New Distribution Concept*, it is not possible to do multithreading, task migration, substitute any values or modify the user interface.

Together with the evaluation of the interaction, there is enough information about the product's usability to discuss its usability quality, but it is not enough when evaluating a multimedia production. In the literature, many authors (Allwood, 1998; Boyle, 1997; Dix *et al.*, 1998; Löwgren, 1993; Newman and Lamming, 1995; Preece *et al.*, 1994 and Shneiderman, 1998) mention that evaluating the usability is not enough. They talk about the productivity, the effectiveness or the functionality of the product, but they do not suggest any method or step in a method to include them. Hence the next step in my own evaluation method.

Table 6.3   Evaluation grounded on principles of the three categories and the four factors perspective

Perspective Categories	Principles/factors	Sandvik Saws and Tools—the New Distribution Concept
Developer's perspective:		
Learning	Predictability	No specific problems.
	Synthesizability	At first problem in building a conceptual model of the system.
	Familiarity	Good, because of the similarities with the web and the use of a business metaphor.
	Generalizability	Good, due to familiarity.
	Consistency	Mostly, but not everywhere, e.g. links to the same pages have different names, some pop-up icon buttons are placed inconsistently.
Flexibility	Dialogue initiative	Mostly the user's, but after every video sequence the control jumps back to the beginning of the video sequence.
	Multithreading	Not possible to do in this production. Suggestion that you can look at the Analytic Toolbox or in the Dictionary at the same time as you look at any Case contents.
	Task migratability	Not relevant in this production because the information is just presented for the user.
	Substitutivity	Not possible. Suggestion to be able to read what is said in the interviews.
	Customizability	The system is unchangeable.

continued

Table 6.3 *(Cont.)*

Perspective Categories	Principles/factors	Sandvik Saws and Tools—the New Distribution Concept
Robustness	Observability	Problems only when you leave the production without noticing.
	Recoverability	No problems due to the functions of the production.
	Responsiveness	Quick enough.
	Task conform-ance	Not changeable by the users.
User's perspective:	Adjustment	No multithreading and no possibility of testing the new knowledge.
	User friendliness	Mostly, but some robustness and flexibility problems exist.
	User acceptance	Probably no problems, but it is not possible to prove the end-user's motivations or feelings without any real end-users.
	User competence	Probably no problems, but there are similar problems as above in grasping the user's understanding of and proficiency in effective interaction with the computer.

6.8   EVALUATION OF THE PRODUCTIVITY

What productivity implies depends on whom you ask. According to Brynjolfsson and Yang (1996) 'Productivity is the fundamental measure of technology's contribution.' and OECD (2001) states that 'Productivity is commonly defined as a ratio of a volume measure of output to a volume measure of input use'. They all appreciate the productivity in quantitative terms, and they appraise the efficiency and effectiveness in working hours or in monetary terms. There are many methods that are available when measuring productivity, but for pricing intangible benefits the PENG method provides support in visualizing the economic value that a particular IT investment adds to an enterprise (Fredriksson, 2003).

Dahlgren *et al.* (2000) have, from their practical experiences, developed the method called PENG ('Prioritering Efter NyttoGrund'; Priorities according to commercial use). PENG is a quantitative oriented method where current utility including 'soft values' can be appraised. All other methods exclude 'soft values' as far as I have noticed. The purpose of the PENG method is to support businesses in identifying both costs and benefits (gross utility) associated with an IT investment. As Fredriksson (2003) writes: 'The PENG method can be used not only for feasibility studies, i.e. appraisals of potential utility, but also for actual economic net value calculations, i.e. appraisals of current utility'.

Preece *et al.* (1994) looked on productivity from the developer's point of view. They state that usability helps to increase product sales and to provide excellent return on investment. The developers are interested in the net utility, which is the difference between the gross utility and the costs associated with the investment. So, when you look at the productivity from a developer's point of view, the PENG method might be useful.

As mentioned earlier, a user might be a customer or an end-user. When thinking of productivity they have different points of view on as to which factors are important to evaluate. The perspectives from both groups are therefore of interest.

A customer is interested in the economic value of an IT investment and when you look at the productivity from a customer's point of view the PENG method might even be useful here. Of course, both developers and customers want useful products, but the factors which ought to be considered are still to be revealed in forthcoming research.

Allwood (1998) looks at productivity from another point of view and asserts that productivity is more then just usability. He looks at productivity from the end-user's point of view, and according to him, the end-user wants the computers to help to improve productivity through efficient

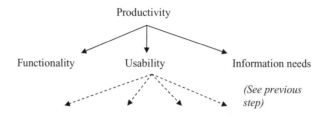

Figure 6.4 Productivity based on Allwood (1998)

functionality. To achieve efficient functionality and higher productivity when fulfilling tasks, programs are needed that include all necessary functionality, usability and information needs (see Figure 6.4). The value on these factors arises in the user's context and in qualitative terms.

Cronholm *et al.* (1999) have also discussed usability. They look at usability in information systems (IS) and state that usability is not enough when designing communication in systems. They argue that *actability* also needs to be considered. They define actability as follows:

> An information system's actability is its ability to perform actions, and to permit, promote and facilitate users to perform their actions both through the system and based on messages from the system, in some business context. The 'degree' of actability possessed by a certain IS is always related to the particular business context.
>
> (Cronholm *et al.*, 1999, p. 4).

They point out the 'relationships between user and task, task and tool, user and tool, as well as between those three and the environment'. Cronholm *et al.* mean that actability and usability are necessary in making systems more actable and usable. This seems to be synonymous with what Allwood (1998) calls productivity. Thus, if the productivity according to him is high then the system's actability is also high.

It is also essential to evaluate the possibility for learning, when you are designing case studies or multimedia applications aimed for learning. Laurillard (1993) has generated a teaching strategy based on the following characteristics:

- *Discursive*—the possibility for user to think step by step. This offers an opportunity for the user successively to build up an understanding of the phenomenon.

- *Adaptive*—the fact that the content of the production changes and adapts to the user's understanding of the phenomenon.

- *Interactive*—the interaction between the user and the production. The user should get meaningful direct feedback on his or her actions, which means that something in the 'world' must change visibly as a result of the user's actions.

- *Reflective*—the production must support the learning process inasmuch as the user connects feedback to his or her actions with the purpose of the task, for instance to link experience to descriptions of experience. The speed of the learning process must be controllable, so that the user can take whatever time he or she needs for reflection.

In trying to evaluate productivity from a holistic perspective, it can be viewed from at least four aspects—the developer's point of view, the customer's point of view, the user's point of view and the teacher's point of view. In Table 6.4 are listed the economic values of productivity evaluated, as well as the productivity from to the end-user's point of view and the teaching strategy factors.

There is only one real problem when looking at the production *Sandvik Saws and Tools—the New Distribution Concept* from the holistic perspective. It is impossible to estimate the product sales and the return on investments, because of lack of information. All other problems either belong to earlier steps in this evaluation method or are deliberately designed limitations.

The two other productions, WM-*data Product Supply and WebDirect—'800 million reasons to use the Internet'* and *Shaping a Virtual Bookstore—the story about Bokus.com*, were not supplemented with any information materials, which also made it impossible to do any economic calculation. Since all three productions have the same strategy according to the learning environments, they all suffer from the same kind of limitations.

This was the last step in the evaluation method and now there is enough information to redesign the product or to release it.

Table 6.4   Evaluation of the productivity

Perspective	Factors	Sandvik Saws and Tools—The New Distribution Concept
Developer's point of view	Economic values	Not possible to do, due to lack of information materials.
Customer's point of view	Economic values	Not possible to do, due to lack of information materials.
End-user's point of view	Functionality	The printing function is not implemented.
	Usability	(Done in an earlier step. See Table 6.3)
	Information needs	Not enough, users need teacher support or textbook.
Teacher's point of view	Discursive	It is possible to some extent, because there is quite a lot of information in the production.
	Adaptive	No adaptation is possible due to the design of the program.
	Interactive	No further interaction about the case, after you have looked at it the first time.
	Reflective	The system gives the user all the time he/she needs.

6.9   CONCLUSIONS

Most researchers stress how important it is to evaluate while designing interactive multimedia, so called formative evaluation. All of them point out that summative evaluation can also be done after implementation. They do categorize and point out different evaluation methods but none discuss *how* to evaluate multimedia production during development and afterwards. There are lots of discussions about how to evaluate usability but, as made clear here, the productivity of a multimedia product depends on more then just the usability. The suggested evaluation method consists of four consecutive steps and they are:

(1) Evaluation of the structure, where the mental construction of the product is analysed.

(2) Evaluation of the interaction, where the dialogue between computer and user is analysed.

(3) Evaluation of the usability, where it is assured how easily a new user would be able to start effective interaction in order to achieve optimum benefit, in how many ways the user can exchange information with the system and how much support the user can obtain to succeed in accomplishing his or her objectives.

(4) Evaluation of the productivity, where the possibility of increasing the quality for the user and/or increasing the profit for the organization of the multimedia product is analysed.

It is of economic interest to know the approximate evaluation time when using this method. Evaluation time for Step 1 depends on how complex the system is. Probably about half an hour is needed. For Step 2 approximately 2 hours, for Step 3 another hour, and for Step 4, 1 or 2 hours depending on how complicated the economic part is and if all the information materials needed are available. The total time for this expert review evaluation, excluding writing a report, is less than a day, which must be regarded as inexpensive and time saving.

The steps in this evaluation method overlap, so being observant in the earlier steps means fewer new problems in the later steps. It is not necessary to follow the whole method at one and the same time. Evaluating the structure can be done as soon as there is anything to evaluate. The next two steps can be evaluated at any time during the development process, and the last step when the multimedia product is completed. If the evaluation is spread out to three or four times during the development process, then both money and time are saved in maintaining the product.

6.10 REFERENCES

Allwood, C.M. (1998) *Människa–datorinteraktion: Ett psykologiskt perspektiv* (Human–Computer Interaction: A Psychological Perspective), Studentlitteratur, Lund.

Bielli, P. and S. Basaglia (2000) 'Multimedia Case Studies: Development and Use in Management Education', in H. R. Hansen, M. Bichler and H. Mahrer (Eds), *ECIS 2000 A Cyberspace Odyssey*, Volume 2, Eighth European Conference in Information Systems, Vienna University, Vienna, pp. 1421–1430.

Boyle, T. (1997) *Design for Multimedia Learning*, Prentice Hall Europe, London.

Brynjolfsson, E. and S. Yang (1996) 'Information Technology and Productivity: A Review of the Literature', *Advances in Computers*, **43**, 179–214, or <http://ebusiness.mit.edu/erik/itp.html>

Cronholm, S., P. J. Ågerfalk and G. Goldkuhl (1999) *From Usability to Actability*, CMTO, Linköping University, Linköping (also published in Proceedings of the Eighth International Conference on Human–Computer Interaction (HCI International'99), 22–27 August 1999, Munich, Germany, Vol. 1, H.-J. Bullinger and J. Ziegler (Eds), Lawrence Erlbaum, New Jersey, pp. 1073–1077.

Dahlgren, L-E., G. Lundgren and L. Stigberg (2000) *Öka nyttan av IT!— Att skapa och värdera nytta i verksamheten med hjälp av PENG* (Increase the Utility of IT! -.To Create and Valuate the Utility in Business with the PENG Method), Ekerlids Förlag, Stockholm.

Dix, A.J., J.E. Finlay, G.D. Abowd and R. Beale (1998) *Human–Computer Interaction*, Second Edition, Prentice Hall Europe, London.

ECCHO (2000) 'ECCH Selected to Distribute European Electronic Cases', *The Newsletter of the European Case Clearing House*, No. 24, pp. 21–22.

England, E. and A. Finney (1999) *Managing Multimedia: Project Management for Interactive Media*, Second Edition, Addison-Wesley, Harlow, UK.

Fredriksson, O. (2003) *Research in Progress Paper: Assessing Economic Net Value from Electronic Commerce-supported Business Process Development*, Third Conference for the Promotion of Research in IT at New Universities and University Colleges in Sweden, 5–7 May 2003, Gotland University, Visby.

Fowler, S. (1998) *GUI Design Handbook,* McGraw-Hill, New York.

Haaken, S. and G.E. Christensen (1999) *Interactive Case Studies— Enablers for Innovative Learning,* Sixth EDINEB Conference 23– 25 June 1999, Bergen, Norway.

Jerkedal, Å. (1999) *Utvärdering—steg för steg* (Evaluation—Step by Step), Nordstedts Juridik AB, Stockholm.

Kristof, R. and A. Satran (1995) *Interactivity by Design: Creating and Communicating with New Media,* Adobe Press, Mountain View.

Laurillard, D. (1993) *Rethinking University Teaching—A Framework for the Effective Use of Educational Technology,* Routledge, London and New York.

Lif, M. (1998) *Adding Usability: Methods for Modelling, User Interface Design and Evaluation,* Dissertation, ACTA Universitatis Upsaliensis, Uppsala.

Lopuck, L. (1996) *Designing Multimedia: A Visual Guide to Multimedia and Online Graphic Design,* Peachpit Press, Berkeley.

Lynch, P. J. and S. Horton (1997) 'Web Style Guide: Basic Design Principles for Creating Web Sites', *Yale C/AIM Web Style Guide,* <http://info.med.yale.edu/caim/manual/contents.html>

Löwgren, J. (1993) *Human–Computer Interaction—What Every System Developer Should Know,* Studentlitteratur, Lund.

Minken, I. and B. Stenseth (1998) *Brukerorientert Programdesign* (User Oriented Program Design), Kirke-, utdannings- og forskningsdep, Oslo.

Newman, W.M. and M.G. Lamming (1995) *Interactive System Design,* Addison-Wesley, Wokingham, UK.

Nielsen, J. (2001) *Ten Usability Heuristics,* <http://www.useit.com/ papers/heuristic>

Nielsen, J. and R.L. Mack (1994) *Usability Inspection Methods,* John Wiley & Sons, New York.

Nielsen, J. and R. Molich (1989) 'Teaching User Interface Design Based on Usability Engineering', *ACM SIGCHI Bulletin,* **21**(1), 45–48.

OECD (2001) *A Guide to the Measurement of Industry-level and Aggregate Productivity Growth*, Paris: OECD Manual, <http://www.oecd.org/pdf/M00018000/M00018189.pdf>

Preece, J. (2000) *Online Communities—Designing Usability, Supporting Sociability*, John Wiley & Sons, Chichester, UK.

Preece, J., Y. Rogers, H. Sharp, D. Benyon, S. Holland and T. Carey (1994) *Human–Computer Interaction*, Addison-Wesley, Wokingham, UK.

Preece, J., Y. Rogers and H. Sharp (2002) *Interaction Design: Beyond Human–Computer Interaction*, John Wiley & Sons, New York.

Seibt, D. (2000) *Final Report BUSINES-LINC Business Innovation Networks—Learning with Interactive Cases,* ESPRIT IV Project No. 26.755, <http://www.wi-im.uni-koeln.de/b-linc>

Shneiderman, B. (1998) *Designing the User Interface—Strategies for Effective Human–Computer Interaction*, Third Edition, Addison-Wesley, Reading, Massachusetts.

Ulfhake, L. (1999) *Utvärdering av multimediaprodukten Sandvik Saws and Tools* (Evaluation of the Multimedia Production Sandvik Saws and Tools), working paper, Esprit project Busines-Linc, Karlstad University, Sweden.

Ulfhake, L. (2001) 'A Model for Systems Development to Produce Pedagogical Software—Multimedia for Business or Pleasure', in Nilsson, A. G. and J. S. Pettersson (Eds), *On Methods for Systems Development in Professional Organisations*, Studentlitteratur, Lund, pp. 142–164.

7

Conceptual Modelling for Creating Multimedia Databases

Lars Erik Axelsson

Information Systems, Karlstad University

7.1 INTRODUCTION

7.1.1 The evolution of conceptual modelling

Data modelling is a general-purpose technique, which can be applied to many problem areas. It is not too risky to assert that data modelling is derived from the early use of database technology. As early as in the beginning of 1960s the need for modelling data and for designing databases was recognized. This was only a few years after the recognition of the need for programming languages (Olle, 1996).

Semantic modelling has been subject of research since the late 1970s. The motivation for this research, according to Date (2000), is that the database systems do not understand what the data in the database

Perspectives on Multimedia R. Burnett, Anna Brunstrom and Anders G. Nilsson
© 2004 John Wiley & Sons, Ltd ISBN: 0-470-86863-5

actually means. The database systems might have a limited understanding of simple data values and perhaps some simple constraints that apply to those values. If nothing more is done to modelling work, all other interpretation of the semantic contents is left to the user. A considerable number of different data models has been developed since the mid 1970s, all of which have addressed the semantic issues from slightly different perspectives (Date, 2000).

Nilsson (1995) emphasizes that data-driven methods for systems development was developed in the middle of 1970s as a counteraction to the function-driven methods from the end of 1960s. These new methods proposed that information systems should be built around more stable concepts (called entities) and the entities need of data as attributes. These new methods were later called data modelling or conceptual modelling.

Preece *et al.* (2002) define a conceptual model as 'a description of the proposed system in terms of a set of integrated ideas and concepts about what it should do, behave and look like, that will be understandable by the users in the manner intended'.

We have already mentioned data modelling, semantic modelling, and conceptual modelling, and presupposed that these are three ways of addressing the same concept. This can of course be a source of confusion. Nevertheless, we can refer to even more related modelling techniques, all of which claim to belong to the group of methods attempting to capture, analyse, prescribe or represent the meaning of the area of interest. Entity-relationship modelling, entity modelling, object modelling, and infological modelling are also often used with reference to the same kind of modelling activity. Doubtless, it is possible to find additional methods that assert that they can perform the same type of activity.

7.1.2   Multimedia databases

Today the area of interest or the Universe of Discourse (UoD) is no longer the same as it used to be. We are not modelling information only of the kind that is common in, for example, a large company or some financial or business profession. Today we want to store more than this in our databases. We want to store pictures, motion pictures, videos, sounds, music and parts of these, and perhaps we want to do it in a way that allows us to formulate questions about impressions, feelings or sense of something. Questions that, for example, sort out pictures that contain pieces of interest (a small part in it that interest us) or let us listen to parts of music that is similar to some tune you have heard somewhere. Databases that can manage tasks of this kind are usually called multimedia databases (Subrahmanian, 1998). Do conceptual modelling methods have the ability to handle these new challenging demands or needs?

7.1.3 Inquiry

The inquiry this chapter is following concerns its kind of problem that might appear when bringing an ISO report on principles on modelling techniques together with two other features, here referred to as multimedia characteristics and emotional factors, when the aim is to create a multimedia database (Figure 7.1). The ISO Report edited by van Griethuysen (1987) is a comprehensive, very accurate and explicit report which covers all aspects of conceptual modelling and conceptual schemas.

The modelling technique principles, according to the ISO report (see Figure 7.1), can be interpreted and categorized as consisting of four essential parts. These parts discuss or describe (1) the notions of conceptualization, (2) general concepts and definitions, (3) constraints, rules or laws, and (4) classification, abstraction, and generalization.

Multimedia can be characterized in several ways. It can for example be described as composed by five working components: integration, hypermedia, immersion, narrativity, and interactivity. *Integration* can be seen as the creation of a hybrid form of expression by combining artistic forms and technology. *Hypermedia* can be explained as associating different media elements to create a personal association. *Immersion* is often explained as being an actor's experience of entering into the simulation or suggestion of a three-dimensional environment, narrativity can be seen as an aesthetic and formal strategy for presenting a non-linear story form and media presentation. *Interactivity* is the ability of the user to manipulate and effect her experience of media (see Burnett, Chapter 1 in this book). In the arrangement illustrated in Figure 7.1, the multimedia characteristics also include all those new media used to contain information. These could be in media formats like video, picture, photo, animation and sound.

The notion of emotional factors in the right-hand box of the figure includes all the aspects that have to be considered in order to describe what the content of the different media formats provides, together with

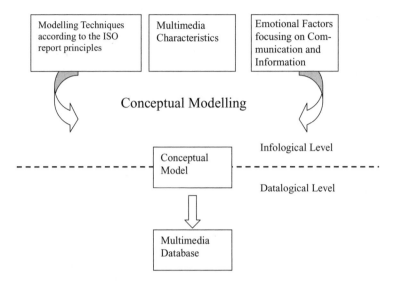

Figure 7.1 What happens when multimedia and conceptual modelling are brought together?

multimedia characteristics. Important aspects to be considered could be impression, feeling, mode or reference to something. According to Figure 7.1, this chapter focuses on communication and information in their reference to these aspects.

When we perform conceptual modelling in a situation like this, that is when the input to the modelling process includes techniques of conceptual modelling, different multimedia characteristics and emotional factors, what problems or new questions do we face? The two curved arrows in the figure symbolize this new conceptual modelling process. The result is, of course, expected to be a conceptual model. The dotted line in the figure distinguishes two levels in development of the information system and these two levels can be seen as the infological and the datalogical perspectives on information system problems, according to Langefors (1995) terminology. The upper part of the figure symbolizes the problem of how to define the information to be made available to the system user. The lower part of the figure then is focused on organizing the set of data in an efficient way in order to implement the system in the multimedia database. This chapter is devoted to the infological issues.

The structure of this chapter is as follows: by way of introduction a description is given of the idea of conceptual modelling starting from the ISO report of van Griethuysen (1987), where fundamental concepts and definitions are considered. Each concept and definition is then discussed, with reference to their capability and/or limitations concerning modelling environments in order to create multimedia databases. This is followed by an analysis of some significant multimedia characteristics.

7.2 THE ISO REPORT

The ISO Report from 1987 on 'Information processing systems— Concepts and terminology for the conceptual schema and the information base', edited by van Griethuysen (1987) is, as mentioned above, a comprehensive, very accurate and explicit report, which covers all aspects of conceptual modelling and conceptual schemas. The next part of the chapter is to a great extent based on the van Griethuysen (1987) report.

One great benefit of the database approach, according to Date (2000), is the fact that the data in the database can be shared.

By sharing the database and the common data in a database, the users communicate with each other through the database management system. The communication exists between the contributors in the system, e.g. between different end-users or between the designer of the system and the end-users. If the communication is expected to be useful there must be a common understanding of the data representing the information. We assume that it is not possible for all the users of a system to meet and discuss the information content of the system and its meaning. Hence, common understanding must be pre-defined and recorded in some external way for all the users (van Griethuysen, 1987).

In other words, traditionally in all conceptual modelling the main precondition is: *all humans that in some way are a part of the UoD share a mental model of this universe of discourse.*

The area this common understanding refers to we call the universe of discourse (UoD), problem area or area of interest. The UoD might be a concrete, solid phenomenon (thing) like an inventory or it might be a more abstract, intangible such as a structure of an enterprise (Boman *et al.*, 1997). Hence the UoD can also be a piece of music or the information a picture is bringing to the users.

The information in the UoD is the part of the world (the system) that is of interest to us. To describe this system we must actually create another system: a data processing system, that contains a linguistic (in one way or another) representation of the first system. We now have two systems of interest: *first the system called universe of discourse (UoD)* and *secondly the system that contains some descriptions of the UoD*. This second system describes or, in other words, models the UoD (van Griethuysen, 1987).

In this chapter we concentrate on the subject of if and how the second system (the describing system) can handle situations where the first system is supposed to contain information in pictures, motion pictures and sounds (e.g. music); information also capable of describing, for example, feelings of calmness in a picture, music that gives impressions of joy or part of motion pictures describing invasion on a beach.

The way conceptual modelling describes or models the UoD according to van Griethuysen, may be summarized as:

- *Conceptualization* and perceiving the UoD as containing objects called entities. Entities are the core notion among others.

- *General concepts and definitions.* Entities are possible to categorize and they can be associated with each other.

- *Constraints, rules or laws* control the behaviour and state of affairs of the entities.

- *Classification, abstraction and generalization* are different ways of structuring information. It is now important to make clear that the ISO report is restricted to expressing only static affairs of the UoD. Van Griethuysen (1987) discusses dynamics but not in a very detailed way in the ISO report.

It is through employing the structures or techniques above that we will create a skeleton description or a conceptual schema of the UoD. This conceptual schema is in fact (or should be) a general agreement about how the users involved perceive the UoD (the shared mental model). It is possible and acceptable that the agreement alters over time or over the life cycles of the system. This description should have a representational form to make it communicable. The representation may have alternative forms (graphic, formal sentences, free text and so on) but regardless of the form chosen, the meaning of the description is of crucial importance.

The ISO report emphasizes that the designer must be free to express the conceptual schema in terms appropriate to the specific UoD of concern and to the user's perception of it. Specifically, no constraints are to be imposed on the entities assumed to exist in the UoD or on the properties they may be asserted to possess (van Griethuysen, 1987).

7.2.1 Conceptualization

In the ISO report, the conceptualization principle is formulated and the report asserts this as a general principle for the conceptual schema.

A conceptual schema should only include conceptually relevant aspects, both static and dynamic, of the universe of discourse, thus excluding all aspects of (external or internal) data representation, physical data organization and access as well as all aspects of particular external user representation such as message formats, data structures, etc.

Conceptualization is a process in which the knowledge of individuals is formed and organized. The results of conceptualization are concepts or more weakly organized conceptions. A concept can be described as identifiable primitives and/or other concepts. Concepts are not classified as entities, attributes or relationships. This kind of classification is a superimposed schema and not an intrinsic feature of knowledge. Obviously, a different kind of classification is possible and depends on the comprehensive goal of the conceptualization process (Kangassalo, 1999).

According to constructivistic philosophy, human individuals develop new knowledge (including concepts) on the basis of previous experiences, from previous knowledge and/or communication with other persons (Preece *et al.*, 2002).

Hence, knowledge and, consequently, constructions of concepts depend on the amount of knowledge an individual possesses. Different individuals will obviously implement the conceptualization process differently even if they started from the same standpoint. We have already asserted above that for conceptual modelling all humans describing the UoD share the same mental model. Obviously this is a statement with some severe limitations. The mental model a certain individual develops or chooses depends to a great degree on the perspective or view of that individual. Probably it is more reasonable to assert that the more complex the concepts, the more likely are the different individuals to reach different results in their conceptualization process. Probably this is an accurate description of a multimedia situation; complex concepts, by definition, exist in multimedia databases.

The result of the conceptualization process must then be transferred and externalized to the external world. In conceptual modelling the first step in this transformation is to create the basic/fundamental building blocks in the conceptual schema from the concepts we have generated in the conceptualization process. Thus the next step is, from a selected part of the world, to force the concepts into an abstract schema.

7.2.2   General concepts and definitions

It is of great importance to provide the modelling process with *a finite set of rules of interpretation*, so that the meaning of the facts asserted unambiguously can be interpreted as true or false (Van Griethuysen, 1987).

That is, if you assert that a picture in the UoD is giving the feeling of calmness, it will be possible to claim that this statement is true or false.

This section contains a brief description of general concepts and definitions in conceptual modelling that serve and support the purpose of a finite set of rules of interpretation (Van Griethuysen, 1987).

- *Entity.* Any concrete or abstract thing of interest, including association among things.

- *Proposition.* A conceivable state of affairs concerning entities about which it is possible to assert or deny that such a state of affairs holds for those entities.

- *Sentence.* A linguistic object, which expresses a proposition.

- *Linguistic object.* A grammatically permissible construct in a language.

- *Term.* A linguistic object that refers to an entity.

- *Predicate.* A linguistic object, analogous to a verb, which says something about an entity or entities to which term(s) in the sentence refer.

- *Name.* A linguistic object that is used only to refer to an entity.

- *Entity World.* A possible collection of entities that are perceived together.

- *Proposition world.* A collection of propositions each of which holds for a given entity world.

- *Universe of Discourse (UoD).* All those entities of interest that have been, are, or ever might be.

- *Necessary proposition.* A proposition asserted to hold for all entity worlds and, therefore, must be part of all possible proposition worlds.

- *Class.* All possible entities in the universe of discourse for which a given proposition holds.

- *Type.* The proposition establishing that an entity is a member of a particular class of entities, implying as well that there is such a class of entities.

- *Instance/Occurrence.* An individual entity, for which a particular type proposition holds, that is, which belongs to a particular class of entities.

- *Conceptual schema.* A consistent collection of sentences expressing the necessary propositions that hold for a universe of discourse.

7.2.3   Constraints and rules or laws

Conceptual modelling is about creating a description or making a prescription of an information system together with its environment and the UoD. Constraints and rules deal with dependencies within and between all of these. Through rules or laws enforced in the method, we define the limitations for what shall be described or modelled in our system. Since the selection of which rules to enforce is to a great extent, arbitrary and voluntary, every system has the possibility of choosing a unique set of laws. The purpose or the intentions of the end-user decide the different constraints of the system and consequently the choice of rules or laws in order to maintain these restrictions or boundaries.

The end-user makes a choice of a certain perspective or view when to analyse the environment and UoD. This perspective or view then governs the selection of rules.

The set of different rules may be structured in many ways. There are static and dynamic rules, local and global rules, impossible and permissible rules and so on. In fact, all kinds of rules may be enforced to support a selected information system. The ISO report discusses static rules or constraints, which verify dependencies between, for example, entities at an instant of time. Dynamic dependencies, in the ISO report, are perceived as governing the evolution of the system.

7.2.4 Classification, aggregation and generalization

Griethuysen (1987) asserts that classification, abstraction, generalization and other rules about UoD is a human process describing a shared mental model of the UoD, but these concepts are not discussed further in the way described below.

According to Batini *et al.* (1992) abstraction is the mental process of selecting some characteristics and properties of an object and excluding some other. We concentrate on properties of a set of objects that we regard essential and forget about their differences. The classification 'abstractions' is used for defining one concept as a class of real-world objects characterized by common properties. It starts from individual items of information (e.g. attributes) (IS-MEMBER-OF). Aggregation abstraction defines a new class from a set of other classes that represent its component parts. We assemble, for example, related field types into groupings of record types (IS-PART-OF). A generalization abstraction defines a subset relationship between the elements of more than one class. It establishes a mapping from the generic class to subset classes (IS-A).

Composition is a variation of aggregation that represents a strong ownership or association between the whole and the parts (Connolly *et al.* 2002).

Obviously we can see all this abstraction as a superimposed set of rules that the end-users agree on. These are all shared mental structures in the shared mental model.

7.3   CONCEPTUAL MODELLING IN A MULTIMEDIA ENVIRONMENT

By way of introduction we must come to an agreement on a definition of multimedia. What is the true sense of multimedia? Apparently there are so many ways of explaining this notion that it is impossible to find a unanimous definition. A synthesis of several longer definitions can be expressed briefly like this:

> Multimedia assume interactivity between the actors in the information process and between users and computer artefacts to a great extent and involve at least two different media.

What kind of problem can arise when conceptual modelling is used to create multimedia databases? The question is actually a variation of a traditional problem of how users understand conceptual models and can be clarified with the help of the classical Norman framework of three types of conceptual models (see Figure 7.2). The figure illustrates the relationship between the design of a conceptual model and a user's understanding of it. 'Designer's Model' is the opinion the designer has of the system. The user has an understanding of the system but if the system image doesn't give a clear picture of the system to the user, the result may be that the user will end up with an incorrect understanding of the system (this is depicted in the upwards arrow to the right in the figure). The 'System Image' illustrates how the system actually works. Do we succeed in creating an ideal world where all three models map onto each other even when we manage a multimedia systems development?

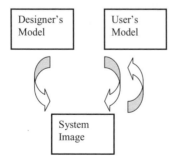

Figure 7.2 Three types of conceptual models (based on Preece *et al.*, 2002).

We can reformulate the question like this:

Is it possible to *conceptualize and classify everything* in the UoD in the sense that actors in an organization can receive plausible answers of questions concerning the semantics of the data stored in the database?

Conceptualise *everything* in the context of multimedia development should now be interpreted as:

- Is it possible to conceptualize information from several different media?

- Is it possible to conceptualize impressions, modes, feelings or references of something?

Analysing the next part of the question *classify everything*, we must explain the following:

- How do we select, define and formulate the properties that we consider to be needed in order to describe the objects or entities of interest in the UoD in a relevant and adequate manner, when they are aimed at conveying impressions, feelings and modes?

7.3.1   The ISO report on principles for modelling techniques

7.3.1.1   Conceptualization of multimedia   Above, we stated that conceptualization implies a process in which knowledge of individuals is formed and organized. Is this process more complicated and more difficult to implement when we are dealing with multimedia connections?

In this context we may often come across the phrase 'multimedia information'. Multimedia information is a strange word, probably misused or not correctly understood. Information of all kinds exists in the part of the world in which we are interested, and when we analyse UoD we will find information in documents, on displays or some individuals will transfer it to us. We find it in pictures, in photos or perhaps in videos. In other words, information exists—and has always done so—in many different media. In fact, information is often conveyed to us in multimedia format. This is nothing new; information analysis has been working in this context for decades. Hence, information exists in many different media at the same time. However, there is no special kind of information named 'multimedia information'.

Does this imply that it is possible to model information of interest regarding these media? Or do we face a new, unique area of problems in conceptual modelling? Apparently the answer must be: this is nothing new, because conceptualization from several different media must obviously already have taken place.

However, has knowledge been formed and organized from impressions, feelings or attitudes, and have these concepts then been transformed to entities? In other words, is it possible to record facts and events about the UoD when we are describing entities like sense of danger, an impression of peace, pictures (or part of pictures) depicting a summer day? Imagine a picture of a many coloured sunset: it can be seen as just a beautiful, romantic sunset or as picture of air pollution where the colours are a token of particles in the air. The picture can also allude to mans' life cycle, with the evening as a symbol for old age. All these perspectives should be covered in the modelling process. Likewise, a picture of birds on wire: they are examples of certain species from the perspective of an ornithologist, but the same picture may give the impression of autumn. The birds are gathered before their voyage to the south. On the other hand, the picture may refer to the songwriter Leonard Cohen and his song *Birds on a Wire*, and refer further to a book about the singer.

If we want to store music, pictures and video in a database, these kinds of consideration must be made. Does it bring new problems to the process of conceptualization or is it evident that knowledge about impressions must also be possible to define. To interpret Kangassalo (2000), if we are human individuals capable of forming and organizing knowledge, we can, or must, be able to conceptualize everything (that does not mean that we always create a valid concept of everything). The new situation that we are confronting through these types of concepts could, however, imply a more difficult task in order to find a common mental model among the users of the system.

Now, to transform concepts of this kind to entities and their properties to attributes is another step in the modelling process. How is this done? Do these kinds of entity (impressions, feelings and so on) have properties? If so, what are the properties of an impression or a feeling? On the other hand, must all entities have properties on the whole? Properties or not in an entity can perhaps not be answered until we know the context of the entities, the users and the question (Axelsson, 2000).

The entities we now discuss are obviously the result of a very subjective judgement in the conceptualization process, resulting in different views or perspectives depending on who is performing the process with what intensions and in which environment and for what purpose? Depending on the background, knowledge, intentions, and context of the individuals who perform the conceptualization, we will probably get different results from the process. Do these circumstances give us a situation where we model a new system every time we find a new perspective or view?

Even Langefors points out reservations of the possibility of enable working communication and he refers to the 'infological equation':

$$I = i\,(D,\,S,\,t)$$

where S is the pre-knowledge and result of the life experience of the individual, so two persons cannot have equal pre-knowledge (S). I is the information or knowledge produced by the information process i applied to the data D and this process is going on during the time interval t.

Any two individuals will receive separate information from the same data or text and they will accordingly, says Langefors, not be able to communicate. However, in some circumstances the communication seems to succeed. Langefors uses, as an example, data processing using a kind of artificial language where objects are identified by means of a pre-arranged identification structure. He calls such a language an object language, and argues that it can be designed as sentences even in natural language and not be restricted to structures used in data processing (Langefors, 1998).

Probably the most crucial problem in conceptual modelling, of multimedia, will appear in the process of conceptualization: the ability to categorize or describe these, by per definition, very subjective concepts in a way that is understood to all actors, in such a way that they can all agree on a common understanding of the conceptual model.

7.3.1.2 General concepts and definitions   As stated above, we find that information exists—and has always done so—in many different media and also in multimedia formats. Information analysis has been working in this context for decades.

After the conceptualization process is completed, then the general concepts are being used to describe the conceptual schemata. These concepts give the opportunity of formulating sentences that express propositions of abstract or concrete things of interest in the universe of discourse. The problem is actually not tied to whether or not the general concepts are usable: they include all the possibilities needed to describe everything that is possible to conceptualize. The major problems are to analyse, to find ways or methods of categorising the very subjective impressions in the conceptualization process; methods to find all the characteristics that are required to understand the UoD.

Consequently, all the definitions and concepts from the ISO report support conceptual modelling for multimedia databases. It is worthwhile to observe that the ISO report does not mention properties as such. There are actually no built-in objections to the general concepts and definitions in conceptual modelling not serving and supporting the purpose of a finite set of rules of interpretation.

7.3.1.3 Constraints and rules or laws   The ISO report discusses mainly static rules or constraints to confirm static dependencies and dynamic dependencies that the report perceives as governing the evolution of the system. There is reason to believe that the complex situations that conceptual modelling for multimedia databases creates increase the need for all kinds of rules to support the systems development.

Abstractions could be understood as set of rules or laws brought in to form structures in the model to help us organize our knowledge about complex entities. Rules can be enforced more or less arbitrarily and voluntarily into the system and should therefore only be looked at as advantageous and not as causing any problem to the modelling process.

7.3.1.4 Classification, aggregation and generalization   Abstractions of different kinds ought to be a useful tool for constructing structures from information in a multimedia context. In these circumstances, we analyse information with complex dependencies from several different media, or when we deal with information of impressions or feelings. The actual purpose of abstractions, according to their definition is, through structuring to obtain simplicity and clarity to complex connections and dependencies between different parts.

On the other hand, when dealing with properties of entities like impressions or feelings, we come across questions such as: do these entities have properties at all? Is for example, 'fear' a generic class of some other feelings such as 'anxiety'?

7.3.2 Multimedia characteristics

Conceptual modelling of a UoD indicates design of a model built from entities that might represent an abstract or concrete thing of interest. The system that is modelled can be a static system but normally there are some ongoing processes taking place.

When the parts and aspects of the real world to be described include multimedia characteristics, such as interactivity, integration, hypermedia, immersion and narrativity, we must be aware of that all these characteristics indicate working processes (see Burnett, Chapter 1 in this book). At the same time they all represent highly subjective or personal phenomena where elements of impression, feeling, mode or reference to something is present. To be able to model these processes (which primarily imply defining entities) we must be able to understand, interpret or conceptualize these phenomena. This calls for an analysis developed from a communication process where intensions and context are essential.

In other words, to be capable of understanding all these processes clearly presupposes an analysis of communication between at least two members taking part in an information exchange. A multimedia context obviously implies exchange of information between actors (like endusers and system designers) and a computer artefact. The analysis of these processes should be directed to achieving a better understanding of the intentions, responsibilities and consequences behind the expressed statements in the communication process.

In conceptual modelling we must analyse communication (and exchange of information) when we want to reflect how changes in the UoD affect the system.

7.3.3 Emotional factors

7.3.3.1 Communication   The topic we are focusing in this chapter is that of actually how to communicate the noticed (percepted) and conceptualized world of one actor to other actors in order to create a mutual understanding. This indicates that the communication process must central to a conceptualization and classification process (which ends in definitions of entities and properties of impressions and feelings in pictures, sounds and so on). To be able to understand each other, communication has the ultimate purpose of sharing intensions regarding the future behaviour and the formulation of norms. An analysis of the interaction process, as stated above, is also a part of the communication process. The semiotic discipline distinguishes several layers in communication, which, at different levels, consider the meaning, form, use and effect of what is communicated (Falkenberg *et al.*, 1998).

- *Physical layer*: the physical appearance, the media and amount of contact available.

- *Empirical layer*: the entropy, variety and equivocation encountered.

- *Syntactical layer*: the language, the structure and the logic used.

- *Semantical layer*: the meaning and validity of what is expressed.

- *Pragmatic layer*: the intentions, responsibilities and consequences behind the expressed statements.

- *Social layer*: the interests, beliefs and commitments shared as a result.

'Intentions', 'communication', and 'negotiations' are some keywords for the pragmatic level. These are words that could have been used to describe features of both the conceptual process and, in particular, the interactivity part in conceptual modelling. We have stressed that significant consideration must be based on the context of the business (or the UoD) and the future purpose (application) of the database and the actors of the database system. We must underline the importance of analysing the environment that the database is intended to serve or support. Obviously in this chapter we are concerned with the pragmatic level of communication.

7.3.3.2 Information about information Conceptualizing impressions, feelings and attitudes is an extremely subjective activity, we have argued. Compared with more traditional decisions in database design as establishing the colour or weight of an object, it is a completely different assessment to appoint the degree of calmness in a picture. However, asserting that colour and weight are important properties for describing an object is undoubtedly a subjective judgement.

Subjective judgements—just as all kinds of judgements—must be able to classify according to some variables. It must be possible to categorize or estimate, for example, the degree of credibility of a judgement. It must be able, in a generic way, to choose a template or pattern of criteria on which to base any kind of judgement. This is just another way of modelling or describing the original (base) modelling done in the first step. We may say that we create and use a meta object system with information about information to describe these new layers of characteristics (Langefors, 1995).

7.4   CONCLUSIONS

The focus in this chapter was upon any questions or problems with conceptual modelling of information in order to create and design multimedia databases.

The multimedia context does not require that the general concepts and definitions in conceptual modelling, suggested in the ISO report, should be abandoned. Information analysis has been working in the context of many different media at the same time for years. Different media like video, pictures, sounds do not hinder us in conceptualising information from these media.

Rules and constraints, as well as abstractions of different kinds as stated in the report, are very useful in conceptual modelling for creating multimedia databases. These concepts have been further analysed since the publication of the ISO report and the new results are highly useful in modelling situations.

There are in fact no problems in analysing and describing concepts originating from different types of media. The real problems arise when when describing impressions, feelings or references to something, provided by *the content of picture, sound,* or *video*. In other words, the question is: Can we create a conceptual model describing these impressions? The crucial problems are the process of conceptualization, and the possibility of categorizing or describing these, by definition, very subjective concepts in a way that is accessible to all actors, so that they all can agree on a common understanding of the conceptual model.

The multimedia characteristics such as *hypermedia* and *interactivity*, and the emotional factors like *communication* (especially the pragmatic and semantic layers of communication) and *information of information*, are all key factors in the process of understanding how conceptual modelling can support multimedia databases.

7.5 REFERENCES

Axelsson, LE. (2000) 'Conceptual Modelling of Multimedia Databases—Is it Possible?', in A.G. Nilsson and J. S. Pettersson (Eds) (2000) *On Methods for Systems Development in Professional Organisations*, Studentlitteratur, Lund.

Batini, C., S. Ceri and S. G. Navathe (1992) *Conceptual Database Design*, Benjamin Cummings, Redwood City, USA.

Boman, M., J.A. Bubenko, P. Johannesson and B. Wangler (1997) *Conceptual Modelling*, Prentice Hall, London.

Connolly, T. and C. Begg (2002) *Database Systems—A Practical Approach to Design, Implementation and Management*, Addison-Wesley, Harlow, UK.

Date, C. (2000) *An Introduction to Database Systems*, Addison-Wesley, Reading, USA.

Falkenberg E. D., W. Hesse, P. Lindgreen, B.E. Nilsson, J.L.H. Oei, C. Rolland, R.K. Stamper, F.J.M. Van Assche, A. A. Verrijn-Stuart and K. Voss (1998) 'A Framework of Information System Concepts', *The FRISCO Report*, Department of Computer Science, University of Leiden, The Netherlands.

Griethuysen, J. J. van (Ed.) (1987) 'Information Processing Systems—Concepts and Terminology for the Conceptual Schema and the Information Base', *ISO/TR 9007*, Switzerland.

Kangassalo, H. (1999) 'Are Global Understanding, Communication, and Information Management in Information Systems Possible?', in Chen, P., J. Akoka, H. Kangassalo and B. Thalheim (Eds) *Conceptual Modeling*, Springer, Berlin.

Langefors, B. (1995) *Essays on Infology*, Studentlitteratur, Lund.

Langefors, B. (1998) 'Information: Data, Knowing, Knowledge, Communication', in P.O. Berg and F. Poulfelt (Eds) (1998) *Ledelselaeren i Norden—En tribut till professor Erik Johnsen, Dafolo*, Fredrikshamn, pp. 84–95.

Nilsson, A.G. (1995) 'Evolution of Methodologies for Information Systems Work—A Historical Perspective', in B. Dahlbom (Ed.) *The Infological Equation—Essays in Honor of Börje Langefors*, Gothenburg Studies in Information Systems, Göteborgs University, Göteborg, Sweden, pp. 251–285.

Olle, T. W. (1996) *Fundamentals of Data and Process Modelling*, Paper presented at the JTC1 workshop in Seattle, USA.

Preece, J., Y. Rogers and H. Sharp (2002) *Interaction Design*, John Wiley & Sons, USA.

Subrahmanian, V.S. (1998) *Principles of Multimedia Database Systems*, Morgan Kaufman, San Francisco.

8

Adding Security to QoS Architectures

Stefan Lindskog
Department of Computer Science, Karlstad University

Erland Jonsson
Department of Computer Engineering, Chalmers University of Technology

8.1 INTRODUCTION

New Internet applications emerge steadily and some of them put new demands on the underlying network. The basic service model provided by the Internet is based on a so-called *best-effort* (datagram) service where data are delivered as quickly as possible, but with no guarantees of timeliness and actual delivery. This service model is unsatisfactory for many new applications, for example, voice over IP (VoIP) and video conferencing.

Users demand specified service levels, which is why the concept of Quality of Service (QoS) has evolved (Wang, 2001; Xiao and Ni, 1999). The basic idea with QoS is to offer additional service classes on top of

Perspectives on Multimedia R. Burnett, Anna Brunstrom and Anders G. Nilsson
© 2004 John Wiley & Sons, Ltd ISBN: 0-470-86863-5

the standard best-effort variant. In a QoS-aware distributed system, a user will be able to choose between various service classes, each with different reliability, predictability and efficiency degrees.

However, until now no security-related service classes have been defined. This implies that users have no chance of configuring their level of security. Security is not recognized as a parameter (or a set of parameters) in current QoS architectures, which is remarkable. One reason for that could simply be that it is difficult to quantify security, something that is needed for security to be treated correctly in such an architecture. Furthermore, security is not a single dimension *per se*, but rather composed by a number of attributes, such as confidentiality, integrity, and availability (the 'CIA'). However, these three attributes describe different, and in many cases contradictory, requirements of the underlying systems and communication channels. This means that two users, both with very high security requirements, could still have very different demands. For example, one user and/or application may require a very high degree of confidentiality, while another user requires a very high integrity level.

This chapter contains a survey of QoS architectures and gives some initial ideas on how QoS architectures can be extended with a security dimension.

In the following, Section 8.2 emphasizes QoS concepts. In Section 8.3, the idea of introducing security in QoS architectures is discussed, while Section 8.4 describes related work. Finally, Section 8.5 concludes the chapter and presents future work.

8.2   AN OVERVIEW OF QOS

No networks are perfect. Every network has its own limitations, such as restrained bandwidth, delays and packet losses. While one application may require high bandwidth and no delays, but can accept a few missed packets during a session, another may settle for low bandwidth, and accept short delays, but no packet is allowed to be missed. Thus, different applications have very different demands on the underlying network as noticed by the Internet community in the beginning of the 1990s, which was when the concept of QoS was born. The basic idea of QoS in a network is to offer a set of service classes to the users, where each class has its own quality parameters as well as quality assurance level. Each user can then choose the service class that best fits her current needs.

8.2.1 Service models

Two fundamentally different approaches to implement QoS have been proposed by the Internet Engineering Task Force (IETF):

- the integrated services (IS) model (Braden *et al.*, 1994);

- the differentiated services (DS) model (Blake *et al.*, 1998).

Both models mentioned above offer additional service classes to the best-effort service offered in the traditional Internet. However, the new classes in IS and DS should be viewed as extensions rather than replacements. Hence, the best-effort variant is still enough for people who only need connectivity, and more advanced and demanding service classes will of course have a higher cost than the basic service.

8.2.2 Integrated services

In the integrated services (IS) model, a signalling protocol for resource reservation is needed, and before any data are transmitted a resource reservation procedure has to take place. The resource reservation protocol (RSVP) (Braden *et al.*, 1997) has been invented for that purpose. The steps in the reservation procedure in RSVP are as follows:

(1) A sender sends a so-called PATH message to the receiver. This message specifies the characteristics of the traffic.

(2) Every router along the path to the final destination forwards the PATH message. For simplicity, we have assumed that all routers along the path support RSVP. However, if that is not the case, tunneling could be used.

(3) When the ultimate receiver receives a PATH message, it responds with a so-called reservation (RESV) message. This message contains information about the requested resources.

(4) When a router on the return path receives a RESV message, it can either accept or reject the request. In case of rejection, an error

message is sent to the receiver and the signalling process is terminated. If, on the other hand, the request is accepted, bandwidth and buffer space are allocated for the flow[†]. Furthermore, the router will keep state information for the flow.

(5) If all intermediate routers accept the RESV message, data transmission can start.

Essentially, the IS model consists of four components: a resource reservation protocol (which could be RSVP, but does not have to be), an admission control routine, a classifier, and a packet scheduler. The admission control component decides whether a resource request could be granted or not. The classifier deduces which class an incoming packet belongs to. Finally, the packet scheduler is responsible for scheduling packets so that they meet their QoS requirements.

In Braden *et al.* (1994), three different service classes are proposed within the IS model, including the standard best-effort service, real-time service, and controlled link sharing.

8.2.3 Differentiated services

The differentiated services (DS) model is very different from the IS model. First of all, the DS model utilizes the type of service (TOS) byte in the IP version 4 (IPv4) header. An application can set three bits in the TOS byte to indicate the need for a certain communication quality, such as low-delay, high throughput, etc. These three bits are today often referred to as the DS field. In the DS model, sets of pre-defined service classes are created using the DS field, i.e., each packet in the network has a certain quality label. Routers within the network will handle a packet differently depending on its assigned service class as given by the DS field. Therefore, the DS model is essentially a relative priority scheme.

Three important questions related to the DS model are: when, how, and by whom is the quality label assigned. A customer who will utilize DS must have a service level agreement (SLA) with her Internet service provider (ISP). An SLA essentially specifies which service classes are

---

[†] The flow abstraction is defined as an distinguishable stream of related datagrams that results from a single user and requires the same QoS.

supported together with the amount of traffic allowed in each class. Two different types of SLAs could be distinguished: static and dynamic SLAs. A static SLA is negotiated on a regular basis (once per month, once per year, etc.), while a dynamic SLA is negotiated once per session. Dynamic SLAs require a resource reservation protocol (e.g., RSVP), which must be used before data traffic could be started. The assignment of the TOS byte, i.e., the quality of service label, is performed when a packet arrives at the ISP.

The DS model requires DS-capable routers. The model can only work properly if (at least most) routers are DS-capable. A DS-incapable router can only forward packets in a best-effort manner.

According to Xiao and Ni (1999), there are a number of advantages with DS over IS. First, the authors claim that DS are much more scalable, because in DS the amount of state information is proportional to the number of classes rather than the number of flows. Second, classification and shaping operations in DS are only necessary at border routers, which most often are not as heavily loaded as the core Internet routers.

8.2.4 QoS parameters

Ideally, a user should have the option to demand exactly a specified service level from a network. Such a specification will typically include a pre-defined set of so-called QoS parameters. Typical QoS parameters are delay, jitter, bandwidth, and loss rate. A fictive QoS specification could then look as follows:

Delay	<400 ms
Jitter	<200 ms
Bandwidth	≥ 64 Kb/s
Loss rate	$<3\%$

Once a specification such as the above has been assessed, a service level request can be given towards a QoS-aware network. This request will either be accepted or rejected by the network. If accepted, a service level agreement between the negotiating parties must have been conducted. If rejected, on the other hand, the network is at the moment either unable or unwilling to offer a communication service according to the request.

8.2.5 Summary

We have in this section introduced the concept of QoS. Two different Internet QoS models, integrated services and differentiated services, have been presented. Furthermore, a short discussion on typical QoS parameters has been given. In the next section, we will continue with a discussion of how QoS architectures could be extended to provide security.

8.3 INTRODUCING SECURITY IN QOS ARCHITECTURES

Nowadays, the importance of security for our society is uncontradicted, and more and more effort is being expended on security-enhancing methods and mechanisms as well as in recovery actions after security breaches. Despite this, there are still no good methods of getting a quantitative assessment of security. It is evident that such a metric will be needed in order to know the level of security for our systems, to compare one system with another, or to compare an improved version of a specific system with the unimproved version. Hence, new ways of measuring security must be invented as a step towards introducing security as a dimension in QoS architectures.

8.3.1 Security definitions

Security is composed of three attributes: confidentiality, integrity, and availability, collectively referred to as the 'CIA'. Confidentiality implies prevention of unauthorized disclosure of information, while integrity means prevention of unauthorized modification of information. Availability means prevention of unauthorized withholding of information or resources. At present, there exist no quantitative measures of confidentiality nor of integrity. Within the fault-tolerant community, availability is defined as follows (Johnson, 1989):

> **Availability** $A(t)$ is a function of time, defined as the probability that a system is operating correctly and is available to perform its functions at the instant of time t.

When this definition was suggested, all introduced faults were assumed to be unintentional and stochastic. However, this does not hold for most of today's operating systems, especially if they are connected to the Internet, since attackers can quite easily set up a (remote) denial of service (DoS) attack against a computer system. Such attacks are definitely intentional and planned activities.

The traditional interpretation of security is that it has two possible states or values, i.e., it is either secure or insecure. We believe, however, that this binary model is both immature and insufficient. Instead security (or rather its attributes) should be regarded as a metric with a whole range of values.

Security definitions and terminology are further addressed in two papers produced by the chapter authors, see Jonsson (1998) and Jonsson *et al.* (1999). In these two papers, additional references to other related work are given.

8.3.2   A note on software metrics

By measuring we might be able to understand our world better, and at the same time interact with our surroundings and improve our lives. According to Finkelstein (1982), Galileo Galilei (1564–1642) once coined the phrase:

'What is not measurable, make measurable.'

The interpretation of this phrase is that science has to find ways to measure attributes of things in which we are interested. In our case, we are interested in defining methods of arriving at quantitative values for the different security aspects to be included in QoS architectures.

Many software engineers and project managers have already realized that software metrics are essential for good software engineering (SE) practice. A whole range of parameters or attributes of things in the SE life cycle have therefore been identified as being measurable or calculable[†]. Today there exist (more or less accepted) measures of the software process, the product as well as the resources used during development. Some examples of measurable or calculable attributes of software are code length, code and algorithm complexity, code structure, reliability, testability, and productivity. However, there is still much work to be done

---

[†] In measurement theory, a measure is a direct quantification, while a calculation is an indirect one.

in this area. Fenton and Pfleeger (1997) provide an in-depth description of this topic.

8.3.3   Security metrics

The discipline of security metrics is, compared with software metrics, much more immature. However, we regard the lack of (good) security metrics as one of the most challenging parts of the research in this area.

Will it ever be possible to quantify security or its attributes? With today's knowledge it seems to be impossible to answer that question. Without comprehensive attempts to find metrics for security, we certainly will not get an answer. It is evident that neither security, nor two of its attributes confidentiality and integrity, are easily measurable; the third, availability, being a possible exception. In fact, we might never be able to measure them directly. What we can and should do, is to try to identify one, or more, indirect measures that could be used as an approximation for a certain security attribute. An example of an indirect measure picked from the SE discipline is Albrecht's function point (Albrecht, 1979). Albrecht's model is an indirect measure that is used to estimate code size by identifying the number of so-called function points in a requirement specification.

Spyropoulou *et al.* (2000) have suggested a concept called 'variant security'. Their hypothesis is that security mechanisms and services are considered to have a security range, and the range is at least binary. Furthermore, they have identified a number of measurable security variables, which could partly be used to quantify a security attribute indirectly. They have, so far, mostly been looking at security variables for confidentiality. Below some examples of such variables are given:

- type of cipher (symmetric or asymmetric);

- key and block length;

- number of encryption rounds (in for example AES[†]).

Here, the assumption is that the more rounds, the better the confidentiality; the longer keys, the better the confidentiality, etc. However, to

---

[†] Advanced Encryption Standard.

approximate an acceptable precise value for one of the three security attributes, more security variables have to be identified.

8.3.4 Security parameters

The idea with QoS is to let users choose their own quality level depending on their current needs. We use the term 'security parameter' to denote an attribute that is available to, and configurable by, the user. Such parameters could either be the same as the security variables discussed in the previous section, or be a combination of two or more of these.

Note, however, that the scale for a security parameter does not necessarily have to be an absolute or ratio scale, where all or most arithmetic analyses are possible on the attribute. See Fenton and Pfleeger (1997) for an in-depth discussion on scales. In some cases, a parameter with a corresponding ordinal scale could be what we are searching for. A 'more-secure-than' relation might be enough in many real situations. An example of such a relation for ciphers could be as follows:

$$AES \geq DES^{\dagger} \geq CAESAR \geq PLAINTEXT$$

The interpretation of this relation is that a message encoded with a CAESAR cipher is harder to analyse (or break) than a corresponding PLAINTEXT message. Similarly, the AES algorithm offers better protection that both the DES algorithm and the CAESAR algorithm.

Before a user starts a communication session he will assign values to the security parameters. For convenience reasons, some form of QoS profile will probably be available. These profiles, which could be one per user, one per role, etc., contain standard or default values of the security parameters for a user, a role, etc. Furthermore, for this scheme to be really useful, the configurable parameters must be understandable by the users. It is also important that users can change the parameters easily, since they will otherwise not change them at all.

Yet another issue with security parameters is the fact that they may change during a session, especially in mobile computing. A user may, for example, have totally different security requirements depending on which network he is connected to. Given a physical movement, both network and service provider may change. A switch from one network

---

† Data Encryption Standard.

to another is called 'roaming' (Lin and Chlamtac, 2001). When roaming takes place, the security requirements may be different. They may, for example, be dependent on the user's willingness to trust network operators other than her own.

8.3.5 Summary

In this section, some security definitions have been given. A note on software metrics was also presented. Security metrics and security parameters were covered in two separate subsections. We stated that the discipline of security metrics is quite immature and must be studied further. It is especially important to find new quantitative measures of security to be able to offer a security dimension to QoS architectures. Security parameters offer security configurability to the users. Some of the issues that should be further studied are those of which parameters should be offered to the users and how are they related to other QoS parameters, how should default user profiles be handled, and should dynamically changing security parameters be supported and what are the implications of such a mechanism.

8.4   FURTHER READINGS

Security is an area that has been studied since the 1960s. The area of QoS, which is not as old as security, has been addressed by researchers and practitioners during the last decade. Security perspectives on QoS have, however, not been studied much. The only work we are aware of is Irvine and coworkers (Irvine and Levin, 1999; Irvine and Levin, 2000a, b; Spyropoulou *et al.*, 2000). However, most of their effort has been on a resource management system (RMS), and especially costing methods have been investigated.

Furthermore, security is sometimes mentioned in the QoS literature. In their research report, Chalmers and Sloman (1998) discuss QoS characteristics in mobile computing environments. In the report, they distinguish between technology-based and user-based QoS characteristics. In the latter, four categories are identified: criticality, perceived QoS, cost, and security. For the security category, four different parameters are presented: confidentiality, integrity, non-repudiation of sending or delivery, and authentication. Unfortunately, the security requirements are not really discussed. Instead, the authors refer to another QoS report (Koistinen and Seetharaman, 1998) and two classical security text books (Pfleeger, 1997; Stallings, 1998). Neither the report nor the text-books discuss the integration of security into QoS architectures further. The fact that security is only mentioned briefly in the QoS literature seems to be the rule rather then the exception, or as Irvine and Levin (2000a) stated:

> Security has long been a gleam in the eye of the QoS community: many QoS RFPs and QoS system-design presentation slides have included a place-holder for security, without defining security as a true QoS dimension.

As previously stated, quantitative measures are needed to express quality levels. Since 1993, research on quantitative modelling and measures of security and dependability have been conducted in the security group at the Department of Computer Engineering, Chalmers University of Technology. The overall goal of the research in this group is to model security within the dependability framework, and in particular to find quantitative measures of security, measures that could be used for predictive purposes. Some publications from the group are Brocklehurst *et al.* (1994), Jonsson *et al.* (1995), and Jonsson and Olovsson (1997).

Finally, it should be noted that common certification standards, such as TCSEC (Trusted Computer Security Evaluation Criteria)—the 'Orange Book', ITSEC (Information Technology Security Evaluation Criteria) and CC (Common Criteria) represent a way of measuring security in the sense that they rank systems into classes and divisions based on design attributes. It could thus be expected that a system from a higher class in general is more secure than a system from a lower class.

8.5 CONCLUDING REMARKS

Security is a very important attribute of modern IT systems. However, there is no good way of making a quantitative assessment of it or of defining service level classes that could be of value to the users and designers. We have made a brief survey of what has been done so far in the area and suggested some potential ways of further progress towards a quality of service concept that would include security aspects.

Still, additional efforts are needed before security (or its attributes) could be used as one or more parameters in QoS architectures. We have to find methods for extending QoS architectures to comprise security. We must also propose definitions of security that reflect the need of the users of the communication channels. Furthermore, we need to define methods of arriving at quantitative values for the different security aspects to be included in quality of service classes or metrics.

8.6   ACKNOWLEDGMENTS

We would like to thank Anna Brunstrom at Karlstad University for helpful comments and discussions related to this work. This research is supported in part by grants from the Knowledge Foundation of Sweden.

8.7   REFERENCES

Albrecht, A. J. (1979) 'Measuring Application Development', *Proceedings of IBM Applications Development Joint SHARE/GUIDE Symposium*, Monterey, California, USA, pp. 83–92.

Blake, S., D. L. Black, M. A. Carlson, E. Davies, Z. Wang and W. Weiss (1998) *RFC 2475*: An Architecture for Differential Services. Status: Informational.

Braden, B., D. Clark and S. Shenker (1994) *RFC 1633*: Integrated Services in the Internet Architecture: An Overview. Status: Informational.

Braden, B., L. Zhang, S. Berson, S. Herzog and S. Jamin (1997) *RFC 2205*: Resource Reservation Protocol (RSVP)—version 1 Functional Specification. Status: Standard.

Brocklehurst, S., B. Littlewood, T. Olovsson and E. Jonsson (1994) 'On Measurement of Operational Security', *Proceedings of the 9th annual IEEE Conference on Computer Assurance (COMPASS'94)*, Gaithersburg, Maryland, USA, pp. 257–266.

Chalmers, D. and M. Sloman (1998) Survey of Quality of Service in Mobile Computing Environments, *Research report 98/10*, Department of Computing, Imperial College, London. Revised February 4, 1999.

Fenton, N. E. and S. L. Pfleeger (1997) *Software Metrics: A Rigorous and Practical Approach*, Second Edition, PWS Publishing Company, Boston, Mass.

Finkelstein, L. (1982) 'What is not Measurable, Make Measurable', *Measurement and Control* **15**, pp. 25–32.

Irvine, C. E. and T. E. Levin (1999) 'Toward a Taxonomy and Costing Method for Security Services', *Proceedings of the 15th Annual Computer Security Applications Conference*, Phenix, Arizona, USA.

Irvine, C. E. and T. E. Levin (2000a) 'Quality of Security Service', *Proceedings of the 2000 New Security Paradigms Workshop*, Ballycotton, County Cork, Ireland, pp. 91–99.

Irvine, C. E. and T. E. Levin (2000b) 'Toward Quality of Security Service in a Resource Management System Benefit Function', *Proceedings of the 2000 Heterogeneous Computing Workshop*, Cancun, Mexico, pp. 133–139.

Johnson, B. W. (1989) *Design and Analysis of Fault Tolerant Digital Systems*, Addison-Wesley, Reading, Mass.

Jonsson, E. (1998) 'An Integrated Framework for Security and Dependability', *Proceedings of the 1998 New Security Paradigms Workshop*, Charlottesville, Virginia, USA, pp. 22–29.

Jonsson, E. and T. Olovsson, (1997) 'A Quantitative Model of the Security Intrusion Process Based on Attacker Behavior', *IEEE Transactions on Software Engineering*, **23**(4), 235–245.

Jonsson, E., M. Andersson and S. Asmussen (1995) 'An Attempt to Quantitative Modelling of Behavioural Security', *Proceedings of the 11th International Information Security Conference (IFIP/SEC'95)*, Cape Town, South Africa, pp. 44–57.

Jonsson, E., L. Strömberg and S. Lindskog (1999) 'On the Functional Relation Between Security and Dependability Impairments', *Proceedings of the 1999 New Security Paradigms Workshop*, Caledon Hills, Ontario, Canada, pp. 104–111.

Koistinen, J. and A. Seetharaman (1998) Worth-based Multi-category Quality-of-service Negotiation in Distributed Object Infrastuctures,

Research report HPL-98-51 (R.1), Software Technology Laboratory, Hewlett-Packard.

Lin, Y. B. and I. Chlamtac (2001) *Wireless and Mobile Network Architectures*, John Wiley & Sons, New York.

Pfleeger, C. P. (1997) *Security in Computing*, Second Edition, Prentice-Hall, Upper Saddle River, NJ.

Spyropoulou, E., T. E. Levin and C. E. Irvine (2000) 'Calculating Costs for Quality of Security Service', *Proceedings of the 16th Annual Computer Security Applications Conference*, New Orleans, Louisiana, USA, pp. 334–343.

Stallings, W. (1998) *Cryptography and Network Security*, Second Edition, Prentice-Hall, Upper Saddle River, NJ.

Wang, Z. (2001) *Internet QoS: Architectures and Mechanisms for Quality of Service*, Morgan Kaufmann Publishers, San Francisco, CA.

Xiao, X. and L. M. Ni (1999) 'Internet QoS: A Big Picture', *IEEE Network*, March/April, pp. 8–18.

9

Partially Reliable Multimedia Transport

Katarina Asplund and Anna Brunstrom
Department of Computer Science, Karlstad University

9.1 INTRODUCTION

Applications that send and receive multimedia data are now becoming increasingly common on the Internet. Many of these applications are today functioning fairly well, at least in certain parts of the Internet. However, as the Internet was originally designed for applications with other types of requirements, the service provided is, in many situations, not nearly adequate. The most important requirements for many of these applications are low delay and jitter. Nevertheless, a limited amount of data loss can often be tolerated, but only up to a certain extent. Above a certain loss rate, the quality degradation will be unacceptable. No service is offered today that provides low delay and at the same time can guarantee a maximum loss rate. The Internet network itself provides a service that is often called *best-effort*. This means that the network does its best to deliver data packets to their destination, but there exists

Perspectives on Multimedia R. Burnett, Anna Brunstrom and Anders G. Nilsson
© 2004 John Wiley & Sons, Ltd ISBN: 0-470-86863-5

no assurance about when, or even if, packets will be delivered. In this situation, a transport protocol becomes responsible for enhancing the service to the application.

Today there exist two standard transport protocols that are in general use on the Internet, namely the Transmission Control Protocol (TCP) (Postel, 1981), and the User Datagram Protocol (UDP) (Postel, 1980). TCP offers full reliability by retransmitting lost data packets. However, this reliability comes at the expense of increased delays and lower throughput. UDP, on the other hand, introduces no extra delay but lacks reliability guarantees. In addition, UDP offers no sequencing for stream data and no flow or congestion control. Neither TCP nor UDP offers a service that is suitable for all multimedia applications. The full reliability provided by TCP is often too costly in terms of delay and throughput, and UDP does not offer any reliability guarantees at all.

In recent years, several proposals for providing adequate service for multimedia applications have emerged. Some proposals involve extending the Internet's service model—the set of delivery services—from the single class of best-effort service to include a wider variety of service classes. Two of these proposals, *Integrated Services* and *Differentiated Services*, were discussed in the previous chapter. Other proposals include building new end-to-end protocols that better support multimedia and/or adapting applications to the service provided by the network. One common approach for designers of Internet-based multimedia tools today is to use UDP and to construct additional reliability services and other functionality as necessary.

Even if new service classes are being deployed in the Internet, it is going to take some time before they are universally offered. Besides, not everyone may want to pay the extra cost that, with no doubt, will entail a better service class. According to this, many applications could benefit from a *partially reliable* service, i.e. a service that does not insist on recovering all, but just some of the packet losses, thus providing higher throughput and lower delay than a reliable transport service. Such a service would be especially beneficial when the channel is lossy and round-trip times are non-negligible (in the order of hundreds of milliseconds), which is often the case for mobile and dial-up telephone links for example.

This chapter takes a look at a transport protocol that provides such a service, the Partially Reliable Transport Protocol (PRTP). PRTP has been developed by the DISCO Research Group at Karlstad University and it is designed primarily for multimedia applications with soft real-time requirements.

The chapter begins with an overview of TCP in Section 9.2, as PRTP is developed as an extension to this transport protocol. The design of PRTP is described in Section 9.3, and in Section 9.4 performance results from experiments where PRTP is used over both an emulated and a real network are presented. The experimental results indicate that transfer times can be significantly reduced when using PRTP as opposed to TCP. This is true when the network loses packets and the application can tolerate a certain amount of packet loss. Section 9.5 gives some suggestions for further readings and Section 9.6 provides some concluding remarks.

9.2 TRANSMISSION CONTROL PROTOCOL—TCP

TCP is the most commonly used transport protocol on the Internet today. It is a connection-oriented, reliable protocol intended for use between hosts in packet-switched networks. TCP guarantees that a byte stream is transferred, error free and ordered, between a pair of processes. Very few assumptions are made as to the service given from the underlying network layer. TCP assumes that the network layer tries its best to route variable-length datagrams to their destinations, but must take into account that the datagrams can be lost, corrupted, duplicated, or delivered in the wrong order. Additionally, TCP must make sure that a receiver, and/or the network, is not swamped with data. Because of the number of different problems TCP has to deal with, it is a rather complex protocol. In the subsequent subsections, a number of the most important mechanisms are discussed.

9.2.1 Error control

A byte stream of data sent on a TCP connection is guaranteed to be delivered reliably and in order at the destination. This is achieved through the use of sequence numbers and acknowledgments. Once the connection is established, data are communicated by the exchange of packets.[†] Each

---

[†] The term *segment* is also used for a TCP data unit.

byte in a packet is assigned a sequence number. Whenever a data packet is sent, a retransmission timer is started. If a positive acknowledgment (ACK) of the data is not received before the timer expires, the packet is retransmitted. The acknowledgment mechanism employed is cumulative so that the acknowledgment number in the ACK indicates the sequence number of the byte expected next. If a packet arrives out of order (i.e. the packet's first sequence number is beyond the sequence number expected next), then the receiver immediately sends a duplicate of the previous sent ACK. The duplicate ACK informs the sender that a packet was received out of order and which sequence number is expected next. If the sender receives three or more duplicate ACKs, then it assumes that the packet has been lost and retransmits it without waiting for the retransmission timer to expire. This mechanism for detecting and retransmitting lost packets is called *fast retransmit*.

9.2.2   Flow and congestion control

TCP is responsible for not sending more data than the receiver or the network can handle. This is achieved by the use of two different 'windows', the receiver window and the congestion window. The receiver window is included in every ACK and it indicates the number of bytes the receiver is currently able to accept beyond the last successfully received packet. The congestion window is an estimate of the number of packets that can be sent into the network without causing congestion. The sender is not allowed to send more data than the minimum of the receiver window and the congestion window. The sender decides the current value of the congestion window by using either the *slow start* or the *congestion avoidance* algorithm. The slow start algorithm, which gives an exponential increase of the congestion window, is used at the beginning of a transfer or after a retransmission timeout. When a connection is established the congestion window is initialized to one[†] packet. The sender starts by transmitting the allowed packet and when the corresponding ACK arrives, the congestion window is increased by one. Next, the sender transmits two packets, and when these are acknowledged, the congestion window is increased by two, one for each ACK. In this way, the

---

[†] The permittable initial congestion window has now been increased to between two and four packets, depending on the packet size.

congestion window proceeds to grow exponentially until the slow start threshold (*ssthresh*) is reached or a packet is lost. If the slow start threshold is reached, the connection leaves slow start and enters congestion avoidance. In this phase, the congestion window is increased by at most one packet during one round-trip time, i.e. a linear increase. Whenever a packet is lost, TCP takes this as a signal of congestion and ssthresh is set to half of the current value of the congestion window. If the packet loss caused a retransmission timeout, the congestion window is set to one packet and slow start is initiated all over again. If, on the other hand, the fast retransmit mechanism was used, the congestion window is set to ssthresh and the connection enters congestion avoidance.

9.3 DESIGN OF PRTP

The partially reliable transport protocol PRTP has been designed for multimedia applications with soft real-time requirements. By providing a partially reliable service, PRTP allows applications to trade a limited amount of data loss in return for higher throughput and lower delay. This section describes different aspects of the design of PRTP. Section 9.3.1 gives an overview of the protocol and Section 9.3.2 describes the mechanism for partial reliability in some more detail.

9.3.1 Overview of design

As mentioned in the introduction, PRTP is not a completely new transport protocol, but has instead been developed as an extension to TCP. This design choice allowed for rapid development of the protocol and made it possible to concentrate on issues directly related to partial reliability. It also means that PRTP can be effectively integrated into the current Internet infrastructure. Since TCP is a fully reliable protocol, some modifications to the reliability mechanisms were of course required. The modifications for PRTP have, however, been limited to the receiver side, where both loss detection and recovery in PRTP are controlled. This allows PRTP-enabled receivers to communicate with regular TCP senders, and thus to take immediate advantage of the large base of legacy servers already available on the Internet. Furthermore, with the receiver-based approach it is the receiver that decides the reliability level desired, based on the demands of the application. This is desirable since it puts the user in control of deciding the trade off between delay and reliability. Aside from the error-control mechanism, no other mechanism in TCP has been modified when developing PRTP.

PRTP allows an application to specify a reliability level between 0 and 100 %, and this specification can be made dynamically at any time during data transfer. The application is then guaranteed that this reliability level will be maintained until a new reliability level is specified, or until the connection terminates. The reliability guarantee that PRTP provides is therefore a deterministic guarantee. If no reliability level is specified, the default value of 100 is used. This means that PRTP provides full reliability and hence standard TCP behaviour.

9.3.2   Partial reliability mechanism

PRTP provides partial reliability by sending 'fake' ACKs for lost packets that are not needed in order to ensure the specified reliability level. As discussed in Section 9.2.1, TCP retransmits all packets that have not been acknowledged before the retransmission timer expires. This time-out and retransmit mechanism often takes extra time, so the transfer time can often be decreased considerably if the number of retransmissions can be minimized.

In order to decide when a 'fake' ACK should be sent, PRTP must first decide when a packet should be considered lost. In contrast to TCP's fast retransmit mechanism, PRTP sees the first arrival of an out-of-sequence packet as a sign of packet loss. When a packet arrives out-of-sequence, the next step is to decide whether the packet is really needed in order to ensure the specified reliability level or if it can be skipped. The decision of whether or not a lost packet is needed depends on the percentage of packets that have arrived successfully up until that moment. However, the decision is also dependent on an *aging factor*. The aging factor makes it possible to give less weight to, or totally disregard packets that are 'old'. For example, an aging factor of 1 means that all packets, new and old, are given the same weight, which is often appropriate for applications such as image transfer. A lower aging factor, on the other hand, means that packets are given lesser and lesser weight as time goes by, something that is often appropriate for applications such as streaming audio and video. If the decision is that a packet is not needed to maintain the reliability level, an ACK for the packet is sent to the receiver in order to avoid a retransmission. If the decision is that the packet is needed, on the other hand, then PRTP uses TCP's standard mechanisms for packet recovery.

One drawback in sending these 'fake' ACKs to the receiver is that congestion control will break down when this happens. That is, when PRTP sends an ACK for a lost packet, the congestion window is not decreased (as it should be), but is instead increased. This property of PRTP makes it too aggressive and TCP-unfriendly in many cases. In order to be a transport protocol for general use on the Internet, PRTP has to be augmented with an alternative congestion control mechanism. The use of explicit congestion notification (ECN) (Ramakrishnan *et al.*, 2001) as a means of signalling congestion in PRTP is explored in Grinnemo and Brunstrom (2001).

9.4 PERFORMANCE EVALUATION

To illustrate the behavior of PRTP, and also how the protocol performs as compared to TCP, this section presents some experimental results based on a Linux implementation of the protocol. Section 9.4.1 shows PRTP's behaviour over an emulated network, and Section 9.4.2 presents experiments where PRTP is tested over a real transatlantic connection, and therefore interacts with regular Internet traffic.

9.4.1 Network emulator experiments

The purpose of the experiments was to evaluate PRTP in networks with different propagation delays and with different degrees of packet loss. The section begins with a description of the experimental setup. Some of the results from the experiments are then presented in a number of graphs, together with an analysis of the data. A more extensive presentation of the results can be found in Asplund *et al.*, (2000).

9.4.1.1 Experimental setup The experimental setup for the experiments is depicted in Figure 9.1. Data were sent between two PCs connected to a 10-Mbps LAN, using ordinary TCP in the sending machine and PRTP

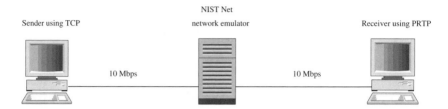

Figure 9.1 Experimental set-up

in the receiving machine. The TCP implementation used in the experiments has two performance improving mechanisms, namely selective acknowledgement (SACK) (Mathis *et al.*, 1996) and timestamps (Jacobson *et al.*, 1992).

SACK enables the receiver to acknowledge data that were received out of order, and at the same time to inform the sender of data that have not yet been received. This information makes it easier for the sender to determine which data should be retransmitted. The timestamps option allows TCP to make more accurate estimates of the round-trip time (RTT), which is used for determining the setting of the retransmission timer.

To be able to vary the propagation delay and the packet loss, a network emulator called NIST Net (NIST, 2000) was used. Using this emulator, the propagation delay was varied between 0 and 500 ms and the packet loss probability was varied from 1 to 10 % in the experiments. Besides these parameters, various values of the reliability level in PRTP were of course tested. For the aging factor two different values, 1 and 0.9, were used. Each run of the experiment consisted of a 5-MB bulk data transfer in order to obtain steady state behaviour. To ensure statistically significant results, each run was replicated at least 20 times, sometimes more.

9.4.1.2 Experimental results   Figures 9.2 and Figure 9.3 show data transfer time as a function of the required reliability level for different packet loss probabilities when the packet delay is set to 50 ms. In Figure 9.2 the aging factor is set to 1, i.e. all packets in the transmission are given the same weight, and in Figure 9.3 the aging factor is set to 0.9, i.e the weights of the received packets are reduced very quickly. As

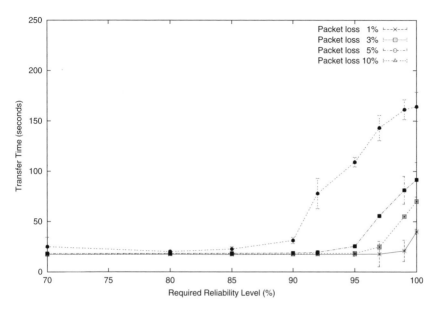

Figure 9.2 Transfer time plotted against required reliability level with packet delay set to 50 ms, aging factor 1.0

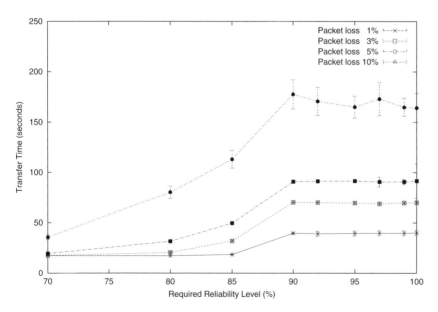

Figure 9.3 Transfer time plotted against required reliability level with packet delay set to 50 ms, aging factor 0.9

can be seen in Figure 9.2, transfer times are constant (and low) as long as the reliability level is so low that retransmissions are not necessary. However, when the reliability level increases beyond the point where PRTP has to begin to retransmit lost packets, transfer times increase very fast. As expected, the figure also shows that the larger the packet loss probability, the greater the increase in transfer time. Figure 9.3 shows a somewhat different picture. As in Figure 9.2, the relative increase in transfer times becomes larger when the packet loss probability becomes higher. However, the increase in transfer times has already stopped at a reliability level of 90 %. An aging factor of less than 1 is primarily meant to be used by applications that want to reduce bursty losses. As the packet loss probability in the experiment is random and not bursty, setting the aging factor to 0.9 has the same net effect on transfer time as adding 10–15 % to the reliability level. As mentioned in the previous section, we have also performed experiments with various other propagation delays between 0 and 500 ms. These experiments show that the relative increase in transfer times generally becomes larger when the propagation delay increases. In summary, the experimental results show that applications that can tolerate some loss could benefit considerably from using PRTP as compared with TCP. Even small loss rates give a noticeable improvement in transfer time, and as the loss rate increases, this improvement becomes larger.[†]

9.4.2 WAN experiments

As part of the evaluation of PRTP, the protocol was also tested over a transatlantic connection. The purpose of these tests was to evaluate how the protocol interacts with regular Internet traffic over a Wide Area Network (WAN). The transatlantic experiments were carried out between the College of William and Mary in Virginia and Karlstad University in Sweden. Data of 2 Mb were repeatedly transferred between a server in Virginia and a client at Karlstad using a set of different values for the required reliability level and the aging factor. The settings of the parameter values were alternated to compensate for the networking load varying over time. Since theses experiments were performed

---

[†] A more elaborate description of the experiments and their results can be found in Asplund *et al.* (2000).

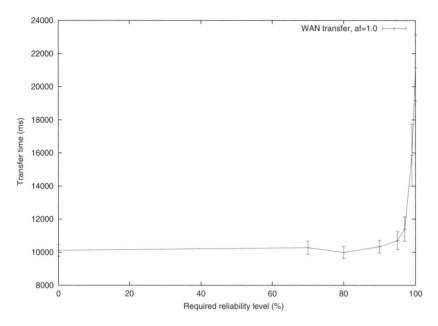

Figure 9.4 Transfer time plotted against required reliability level with WAN Experiment

over a live network it was not possible directly to control the network load or packet loss rate. SACK and timestamps were used also in these tests.

Figure 9.4 shows the obtained results for an aging factor of 1.0. As can be seen from the figure, the transfer time is approximately cut in half if some packet loss is allowed. Only a small reduction, below 100 %, is needed in the required reliability level to achieve this speed up. The average values displayed in Figure 9.4 do not display the dependence of the results on varying network load. Figure 9.5 shows the results for each one of the sample runs at full reliability and at 97 % reliability. We can see from Figure 9.5 that the network load was higher at the beginning of the experiment. As expected, the difference between the two reliability levels appears to be larger when the network load is high.

Over the given transatlantic connection there was enough congestion to lead to packet loss and degraded performance. In this situation, the use

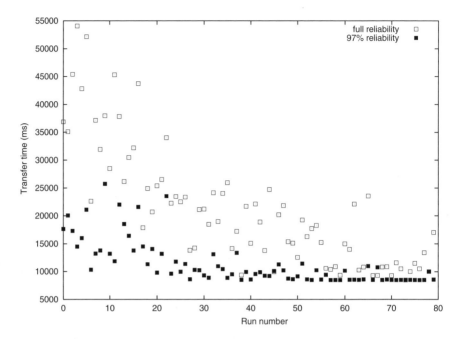

Figure 9.5 Transfer time for individual runs

of PRTP led to a considerable speeding up. It should be noted, however, that this speeding up is due not only to the use of partial reliability, but also to the use of a more agressive congestion control.

9.5   FURTHER READING

Several transport protocols that in different ways provide a partially reliable service have been proposed. One example is POCv2 (Conrad *et al.*, 1996) which in addition to partial reliability also provides a partially ordered service, i.e. a service where not all of the transmitted data must neccessarily be delivered in sequence. POCv2 divides application data into a number of messages that are each assigned a reliability class. The reliability classes in POCv2 are reliable, unreliable and partially reliable. Reliable indicates that messages are retransmitted until they arrive successfully and unreliable that messages are never retransmitted. Partially reliable messages are retransmitted as long as subsequent messages are not delayed, and the receiver has other messages to deliver to the application.

The Partially Error-Controlled Connection (PECC) (Dempsey, 1994) is a partially reliable extension to the general purpose transport protocol Xpress Transfer Protocol (XTP) (Strayer, 1995). In PECC, applications specify their loss tolerance and PECC tries to respect this tolerance. However, lost packets are only retransmitted when they will not add additional delay to the data delivery. This means that the loss tolerance cannot always be respected, and that PECC does not provide a deterministic reliability guarantee to the application.

The Stream Control Transmission Protocol (SCTP) (Stewart, 2000) is a transport protocol developed by the IETF which is primarily aimed for transportation of telephony signalling messages. However, SCTP can also be used for other kinds of data transport. A partially reliable extension to SCTP has been proposed, called PR-SCTP (Stewart *et al.*, 2003). The extension allows the sender to decide whether a packet should be retransmitted or not. If the decision is not to retransmit, the sender sends a message that tells the receiver not to wait for this packet (or packets with lower sequence numbers) any longer.

For those who are interested in reading more about partially reliable transport protocols, a taxonomy and survey of retransmission based, partially reliable transport protocols can be found in Grinnemo *et al.*, (2002).

Several other transport protocols aimed at multimedia applications, but which do not provide a partially reliable service, have also been proposed. One of these is the Datagram Congestion Control Protocol (DCCP) (Kohler *et al.*, 2003), which provides an unreliable flow of congestion-controlled datagrams. The protocol is intended for applications such as streaming media that require TCP's flow-based semantics but does not want the overhead of TCP's full reliability.

A protocol, which is used by many multimedia applications today, is the Real-Time Transport Protocol (RTP) (Schulzrinne *et al.*, 1996). RTP provides a number of end-to-end services that are useful for many multimedia applications. These services include payload type identification, sequence numbering and time stamping. However, RTP itself does not provide full transport functionality, but is designed to work together with another transport protocol (typically UDP). Furthermore, RTP does not guarantee packet delivery or that packets are delivered in sequence, nor does it ensure timely delivery.

9.6   CONCLUDING REMARKS

As an example of a protocol that provides extended services for multimedia applications, we have in this chapter described the partially reliable transport protocol PRTP. PRTP is a transport protocol especially aimed at applications with soft real-time requirements, and it provides applications with an easy to use and very flexible partially reliable transport service. PRTP has been implemented as an extension to TCP with all necessary modifications localized to the receiver. These design choices allow PRTP to be integrated easily into the current Internet infrastructure.

To illustrate the behaviour of PRTP, various experimental results are presented. The experimental results suggest that PRTP can indeed decrease transfer times in networks where packet loss occurs. This is especially so for applications that can tolerate rather high loss rates (5–10 %), and when the average loss rate in the network is slightly below the limit, the performance improvement is considerable.

The next chapter describes another transport protocol aimed for multimedia transport, called TCP-L, which allows packets with bit errors to be delivered to the application. TCP-L is especially intended for communication over wireless links.

As discussed in the previous section, several different transport protocols have been proposed for multimedia applications. It is today difficult to say which of these protocols (if any) is going to be in general use in the future. Only time can tell how the Internet will evolve better to support these applications.

9.7   REFERENCES

Asplund, K., J. Garcia, A. Brunstrom and S. Schneyer (2000) 'Decreasing Transfer Delay Through Partial Reliability', *Proceedings of PROMS 2000*, Cracow, Poland.

Conrad, P. T., E. Golden, P. D. Amer and R. Marasli (1996) 'A Multimedia Document Retrieval System Using Partially Ordered/Partially Reliable Transport Service', *Proceedings of Multimedia Computing and Networking*, San Jose, CA.

Dempsey, B. J (1994) Retransmission-based Error Control for Continuous Media, *PhD thesis*, University of Virginia.

Grinnemo, K.-J. and A. Brunstrom (2001) 'Evaluation of the QoS Offered by PRTP-ECN—a TCP-compliant Partially Reliable Protocol', *Proceedings of IWQoS 2001*, Karlsruhe, Germany, pp. 217–231.

Grinnemo, K.-J., A. Brunstrom and J. Garcia (2002) A Taxonomy and Survey of Retransmission Based partially reliable Transport Protocols, *Karlstad University Studies 2002:34*, Dept. of Computer Science, Karlstad University.

Jacobson, V., R. Braden and D. Borman (1992) *RFC 1323*: TCP Extensions for High Performance.

Kohler, E., M. Handley, S. Floyd and J. Padhye (2003) 'Datagram Congestion Control Protocol (DCCP)', *Internet Draft*, draft-ietf-dccp-spec-03.txt, *work in progress*.

Mathis, M., J. Mahdavi, S. Floyd and A. Romanow (1996) *RFC 2018*: TCP Selective Acknowledgment Options.

NIST (2000) 'NISTNet Network Emulation Package', <http://www.antd.nist.gov/tools/nistnet/> October 6, 2003.

Postel, J. (1980) *RFC 768*: User Datagram Protocol.

Postel, J. (1981) *RFC 793*: Transmission Control Protocol.

Ramakrishnan, K., S. Floyd and D. Black (2001) *RFC 3168*: The Addition of Explicit Congestion Notification (ECN) to IP.

Schulzrinne, H., S. Casner, R. Frederick and V. Jacobson (1996) *RFC 1889*: RTP: A Transport Protocol for Real-time Applications.

Stewart, R. (2000) *RFC 2960*: Stream Control Transmission Protocol.

Stewart, R., M. Ramalho, Q. Xie, M. Tuexen and P. Conrad (2003) 'SCTP Partial Reliability Extension', *Internet Draft*, draft-ietf-tsvwg-prsctp-00.txt,*work in progress*.

Strayer, W. (1995) Xpress Transport Protocol 4.0 Specification, *XTP Forum Inc*.

10

Bit Error Tolerant Multimedia Transport

Stefan Alfredsson and Anna Brunstrom
Department of Computer Science, Karlstad University

10.1 INTRODUCTION

The Internet has traditionally been used to transport content that is sensitive to errors. As the Internet was designed for a wide range of heterogenous networks, the transport protocol used must take measures to ensure that the information is intact. It is easy to envisage that an error in executable content might cause it to malfunction at the time of execution. Even minor errors in an e-mail could change the semantics of the message.

However, the characteristics of the Internet user base is changing. From being used mostly by military and academia, the major user base has shifted to private consumers. A driving force is entertainment, in the form of music and video. This data can be summarized as multimedia content.

Perspectives on Multimedia R. Burnett, Anna Brunstrom and Anders G. Nilsson
© 2004 John Wiley & Sons, Ltd ISBN: 0-470-86863-5

Multimedia content differs from traditional content in a few ways. Due to human perception, small errors in the presentation of sound or pictures can be accepted because the ear or eye will regard it as 'noise' and compensate with help of perceived correct information. Traditional content, for example a computer program, can not accept such errors because the computer assumes that all instructions are correct.

Another difference is in the characteristics of transport. A distinction can be made between bulk data transfer and 'real time' data transfer. Bulk transfer is usually not sensitive to delays or jitter (differing interpacket delays). The transfer rate may be allowed to fluctuate quite a lot, without any impact. Real-time transfers, on the other hand, require that data are delivered at the expected time, or it will not be useful anymore. For example, if a video frame arrives too late, it can not be used anymore since its position in the video stream has already passed.

Another area of consumer interest is mobility. Mobility is often achieved by using wireless communication. The user has a mobile terminal, which communicates with a base station. The most common example is the cellular phone, but as the third and fourth generation mobile systems appear, the border between telephony and data services is being erased.

Given this information, it can be seen that multimedia content (a) can accept errors, and (b) has greater demands on the transport characteristics in real time situations. Therefore, it could be beneficial to trade errors for better network performance. Wireless radio networks are a problematic environment for Internet communications, where such a trade off could be beneficiary. A wireless network exhibits higher error rates and greater delay variations. The underlying radio transmissions experience fast variations in received signal strength (fading), and the radio waves bounce on objects such as buildings (multi-path effects). The received signal strength is also weakened by the distance between a sender and receiver. User mobility therefore leads to additional variations in received signal strength.

Communication over the Internet often uses the TCP (Postel, 1981b) protocol for transporting data. As more and more multimedia applications are being used on mobile terminals, the possibility of improving the transport protocol performance by modifying TCP to allow bit errors is investigated in this chapter. This is similar to the PRTP protocol described in the previous chapter, in that reliability is traded for performance. PRTP can be said to work on a packet loss basis in wired or wireless networks, while the approach presented in this chapter is aimed at bit errors in wireless networks.

The rest of this chapter is structured as follows. Section 10.2 contains a background on transport protocols, more specifically TCP and the bit-error tolerant modification TCP-L. A performance evaluation that compares the performance of TCP-L to the performance of TCP has been performed. Section 10.3 presents the experiment set-up and implementation, while Section 10.4 presents the experimental results. Section 10.5 discusses some further readings, and finally, Section 10.6 gives some concluding remarks.

10.2   TRANSPORT PROTOCOL BACKGROUND

This section gives an introduction to TCP, and the modifications introduced by TCP-L. TCP is introduced in the previous chapter, but is presented here as well to make the chapter self contained.

10.2.1   TCP

As previously stated, TCP is the most used transport protocol on the Internet today. It was deployed on the Internet at the beginning of the 1980s, as the then used protocols did not scale well (Postel, 1981c).

The purpose of TCP is to provide a reliable and ordered transport service. What is transmitted at the sender should be received identically at the destination. TCP works by dividing a stream of data into smaller packets (also called segments), typically 576 to 1500 bytes. These packets are encapsulated in IP (Postel 1981a) packets, marked with the address of the sender and receiver, and sent on the network. If the receiver is not on the same local network, intermediate nodes (routers) take care of forwarding the packets to the correct destination by looking at the address tags. Once received, the header information is stripped, and the TCP packets are assembled to form a stream of data again.

To achieve reliability and ordered delivery of data, sequence numbers are used. Each byte is assigned a sequence number, counted from an initial arbitrary offset (initial sequence number). The sequence number of the first byte in a packet is included in the TCP header. The sequence numbers of packets that are received correctly are used to acknowledge the data to the sender. If a packet or acknowledgement is lost, the sender will then retransmit the unacknowledged data after a timeout. When packet loss occurs this may lead to packets arriving out of order at the receiver. Packets may also be reordered when forwarded over different paths within the network. The sequence numbers are thus also used to ensure that the data are delivered to the application in the right order.

Another aspect of reliability is the integrity of data. Integrity is ensured by a checksum that is computed at transmission and verified at reception. Packets with a correct checksum are acknowledged, while packets with an incorrect checksum are discarded. The sender will then retransmit the unacknowledged packets, as in the case of packet loss.[†]

To make TCP use the network it operates upon efficiently, several packets can be in flight at the same time. Instead of sending one packet, awaiting the acknowledgement and then sending the next packet, a sliding window technique is used. The receiver advertises the amount of data it can handle, called the receiver window, to the sender. The receiver window tells the sender how much data it can send before an acknowledgement is needed. This avoids overloading the resources at the receiver.

TCP must also be careful with the resources of the network. Even though the receiver may be able to handle more data, the network can be overloaded. For example, many data sources may try to send data through a specific router at high speed. If the output link of the router cannot handle this amount of data, it is buffered, awaiting an available time slot for delivery. When the buffer is full, the router must drop the new packets that arrive[‡] until there is more buffer space available. This state is called congestion (Allman *et al.*, 1999). The solution of TCP for getting out of this state is to lower the transmission rates of the senders. This is done by using a congestion window that acts as a second limiting factor. The sender may not transmit more data than what fits in the receiver or congestion window, whichever is smaller.

[†] On a side note, there are rare cases when the checksum fails to indicate packet errors, as noted in Stone and Partridge (2000). This means that even if the checksum is correct, there could still be errors in the data.

[‡] There are several approaches when dealing with a full buffer, dropping the newly incoming packet being one of them.

In the beginning of a session or after a timeout, the congestion window is set to one packet.[§] The window is increased when acknowledgements are received, to allow for more data to be injected into the network. This algorithm is called slow start. If slow start is frequently initiated, this can severely affect TCP throughput.

In cases of slight congestion, the fast retransmit algorithm is used. When a packet is lost, the sender would normally detect this through a timeout waiting for an acknowledgement. Alternatively, if packets are still flowing through the network, a packet loss will result in a gap in the data stream (i.e. packets arrive out of order). When packets arrive out of order, the receiver sends duplicate acknowledgements requesting the same missing data. The sender uses these acknowledgements for early detection of packet loss instead of waiting for the timeout to occur. Fast retransmit then immediately retransmits the missing packet. The congestion window is reduced, but less aggressively than after a timeout.

10.2.2   TCP over wireless links

Wireless links behave differently from wireline links constructed of, for example, fiber or copper. Due to the properties of radio-based bearers, errors are more prone to appear in this environment. Multipath effects, fading and path loss can make it harder to separate the signal from noise. To improve the performance of wireless links, specialized link layers for the problematic environment are used. These typically divide incoming packets into link-layer frames, much smaller than the packets themselves. The link layers then utilize their own retransmission scheme for retransmitting frames that have been damaged or lost. Schemes such as interleaving and forward error correction are also used to increase performance. If a link layer fails to deliver a link-layer frame, the whole TCP packet is often discarded. If the link layer allows damaged frames, the assembled TCP packet will contain errors.

A TCP packet travelling over a wireless link therefore has a higher probability of being lost or corrupted than one transmitted over a guided medium. In case of links that propagate erroneous frames, the receiver will detect that the checksum does not match the content of the packet and discard it. This means that from the senders perspective, it will not

---

[§] A slightly larger initial congestion window is allowed.

be able to tell the difference between a packet dropped in a router due to congestion, a packet dropped at the wireless link, or a packet dropped at the receiver due to an invalid checksum (Biaz and Vaidya, 1997). In all cases, the sender will perform its congestion avoidance algorithm, even though it is only needed in the first case.

10.2.3 TCP- L

One conclusion to be drawn from the last two sections is that if the different causes of packet loss could be distinguished, the performance of TCP could be improved. How can this be done? The link-layer can send a notification to the sender (Ding and Jamalipour, 2001). Another method is to have the receiver indicate that it indeed has received the packet, but it was damaged (Balan *et al.*, 2002; Garcia and Brunstrom, 2002). The sender retransmits the particular packet and continues the transmission without decreasing the transfer rate. A third method, explored in this chapter, is for the TCP receiver to accept that the packet contains errors, and make the best of the situation. Comparing these approaches, it is seen that in the first two cases modifications must be done at both link (or receiver) and sender. The latter approach only requires modification at the receiver. Since no retransmission is done, the link utilization have a better chance of being optimal as (i) the congestion avoidance algorithm is not started and (ii) the packet is only sent once. The trade off is that the application must be able to accept errors in the data it receives. Good candidates for this is multimedia applications where the final 'consumer' of the data is a human. Minor errors in images and audio will be concealed by the senses of seeing and hearing.

The idea of accepting errors in TCP at the receiver is called 'TCP-L' (formerly TCP Lite), and was first explored by Alfredsson (2001). The name reflects a more lightweight version of TCP, in the sense that it does not always perform retransmissions of corrupt packets. When a packet arrives at the host, its checksum is verified. If it is correct, the packet is processed as normal. If the checksum is incorrect, then header recovery is performed. The header contains the sequence number which is sensitive to errors, along with other information that is less sensitive. By making some assumptions of the transport protocol usage, most parts of the header can be reconstructed to safe values. After the header has been recovered, it is passed on in the stack as if the checksum was correct.

It is then up to the application to handle the errors that occur in the data stream. A detailed description of the header recovery can be found in Alfredsson and Brunstrom (2003).

10.3 PERFORMANCE EVALUATION

Experiments were done to get an indication of how TCP and TCP-L behave under varying wireless link conditions.

The experiments were performed by using three networked computers, as illustrated in Figure 10.1. The first machine acted as the sender running the Linux 2.4.20 kernel, the second machine emulated a wireless link via the FreeBSD 4.5 dummynet (Rizzo, 1997) network emulator, and the third machine acted as the receiver running the TCP-L modification implemented in Linux 2.4.20. The experiments consisted of sending bulks of data from the sender to the receiver. The traffic between the sender and the receiver was routed through the emulator, which caused errors in the passing packets based on the state of the emulated radio channel. The emulator also introduced delay and bandwidth limitations.

In the experiments, the link was modelled by a bandwidth and a delay parameter, that resembled the UMTS (Muratore, 2000) third-generation mobile system. A bandwidth of 384 kbit/s and a 70 ms end-to-end delay were chosen as typical parameters of such a system.

Two error models were used. In the first experiments random independent errors were used. The probability of error ranged from 0 to $2.5*10e-5$. In the second set of experiments, the amount of errors was fixed at 10e-5 bits. Instead of varying the amount of errors, the amount

Sender   Wireless link emulator   Receiver using TCP–L

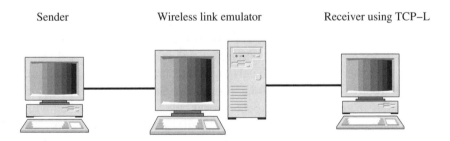

Figure 10.1 Experimental set-up

of burstiness changed. The distribution of errors ranged from random placement to becoming more and more grouped together in bursts.

Two different packet sizes were used (to see if packet size had an effect on performance). By specifying the maximum transmission unit at the sender, packets of 576 bytes and 1500 bytes were transmitted.

To gather statistics, the network traffic was captured at the receiver using the application tcpdump. These captures were then analysed with tcptrace (Ostermann, 2003), which produces statistics on individual TCP connections. Tcptrace can, for example, report the throughput, the number of retransmitted packets and the total number of packets of the connection. These data were then extracted and statistically processed to produce the graphs shown in the next section.

10.4 EXPERIMENT RESULTS

10.4.1 Randomly distributed errors

Figures 10.2 and 10.3 show a comparison between TCP and TCP-L in the presence of errors. The amount of errors are shown on the y-axis, while the throughput is shown on the x-axis. The throughput at zero errors indicates the upper bound of the throughput capacity at the given packet size. As more and more errors are introduced, the throughput is reduced to a varying degree. TCP, the lower curve in both graphs, reacts drastically to increasing amounts of errors. These errors translate to packet loss. The sender uses this as an indication of congestion, and lowers the sending rate. TCP-L, the upper curve, recovers well from most of the errors. As the error rate increases, slightly fewer packets can be recovered. This leads to a minor reduction in throughput, compared with the reduction that regular TCP experiences.

Another aspect that can be seen from the two figures is the impact of packet size. In Figure 10.2, a packet size of 576 bytes is used. This means that there is more overhead for packet meta information (headers), which leads to a reduced optimal throughput. However, as the error rate increases, there is less chance of a packet being in error compared with the larger packet size in Figure 10.3. Also, for the smaller packet size less bandwidth is wasted when retransmissions occur, compared with retransmissions with the larger packet size.

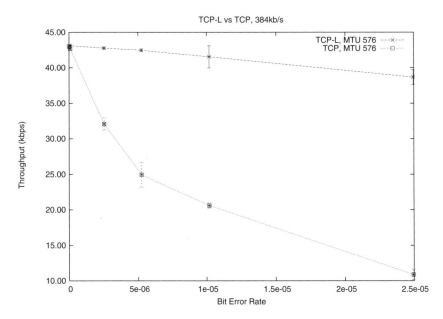

Figure 10.2 Increasing amount of randomly distributed errors, MTU 576

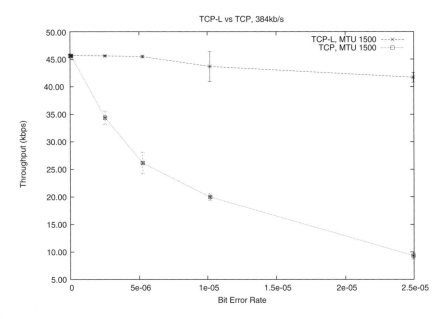

Figure 10.3 Increasing amount of randomly distributed errors, MTU 1500

10.4.2 Impact of burstiness

The above experiments assumed that all errors were randomly distributed. Depending on the link technology, errors may appear in bursts. The impact of error burstiness was explored by using a fixed amount of errors, and then varying the error distribution. In Figures 10.4 and 10.5, the *y*-axis shows the throughput as before. The *x*-axis shows the average probability that errors are placed near each other. At 0 %, the errors are randomly placed. This result can be seen to correspond well with the other experiments at the same bit error rate (10e-5). As the probability then increases, the rate of error burstiness increases as bit errors become more and more grouped together.

In Figure 10.4 the graph shows that as bursts become longer, throughput for TCP is improved. The reason for this is that there is no difference if there is one or more errors in a packet, it will be discarded anyway, but with a fixed amount of errors, and these errors grouped together, there

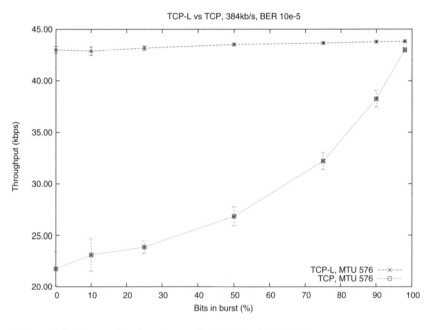

Figure 10.4 Increasing burstiness, fixed BER. MTU 576

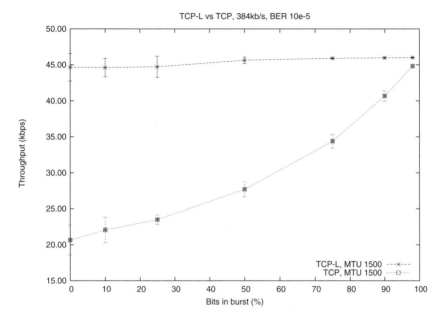

Figure 10.5 Increasing burstiness, fixed BER. MTU 1500

will be fewer packets overall that are damaged. With a lower packet loss comes better performance, as seen in the previous experiments.

TCP-L maintains a throughput near the optimal, independent of the burst distribution. A small improvement is made as bits become more and more grouped together. Before the experiments were done, it was unclear how the bursts would affect the performance of TCP-L. The main question was if the longer bursts would make error recovery harder, compared to recovery from single bit errors. However, as seen in the figures, TCP-L also improves its throughput when there are longer bursts.

The experiments in Figure 10.5 use the larger packet sizes. As for randomly distributed errors, the impact of lower overhead for larger packets is seen.

10.4.3 Experiment conclusions

The presented observations suggest that TCP-L can achieve higher throughput compared with regular TCP. When randomly independent bit errors are introduced, TCP-L is negligibly affected. Regular TCP performs very poorly, with halved throughput at a residual bit error rate of 10e-5.

For the case of a fixed amount of errors and varying error distribution, TCP-L also performs well. The throughput difference between random and bursty errors is negligible. As errors are grouped more and more into bursts, the throughput of TCP increases. This is because larger bursts result in fewer packet losses, which in turn translates into an increase in throughput.

Some aspects that will be further investigated include the impact of link-level retransmissions and coding mechanisms on the residual error rate. The error tolerance of applications is dependent upon content encoding and compression, which should also be investigated.

10.5   FURTHER READINGS

TCP-L is one proposal for increasing the performance of multimedia applications over networks that experiences higher error rates. A similar protocol that provides a progressively reliable transport service for interactive wireless multimedia is Leaky ARQ, presented in Han and Messerschmitt (1999). Leaky ARQ allows for delivery of erroneous packets, and selective retransmission of packets. This means that the application may be content with the data it receives, even if it contains errors. If there are important data that are damaged, the application can choose for it to be retransmitted.

For interactive streaming multimedia applications, UDP (Postel, 1980) is an often used protocol. UDP, like TCP, utilizes a checksum to detect errors in the payload. In Larzon *et al.* (1999), a modification of UDP known as UDP Lite is described. It supports the delivery of errors in packet payloads, by giving the possibility of defining the checksum coverage. The UDP header has a length field that normally indicates the size of the payload. This is redefined in UDP Lite to instead indicate the amount of data that the checksum covers. Some applications that use UDP use their own header information within the UDP payload. This header can then be protected by the checksum, while subsequent data can contain errors. However, the protocol does not allow for error detection in the unprotected part of the packet.

Several other approaches propose improving TCP performance in unreliable wireless networks while keeping the paradigm of reliable transport. Based on their placement in the network, they can coarsely be categorized into *link-layer, split connection* and *end-to-end* enhancements.

Link-layer modifications only require modifications to the network component. The link layer can be made as reliable as possible by using local retransmissions until successful. This can, however, incur large packet delay variations, and cause bad interactions with transport layer retransmissions. Instead, the link layer may give up after a limited number of retransmissions and drop the packet. Other link layer modifications are based on the awareness of the packets traversing the link. For example, one well known technique for handling TCP packet loss in the link layer is Snoop (Balakrishnan *et al.*, 1995). Snoop uses an agent located at the base station that monitors traffic from and to the mobile terminal. Packets going to the mobile terminal are cached until an acknowledgement has been seen. When snoop detects that packets are lost on the wireless part, it does a local retransmission of the cached packet. The sender, therefore, never detects the packet loss. Further link layer proposals include WTCP (Ratnam and Matta, 1998), and TULIP (Parsa and Garcia-Luna-Aceves, 2000).

Split connection approaches treat mobile networks specially, by using a separate transport layer connection between the mobile terminal and the base station. The base station then communicates with the content server over the fixed network, using TCP. Thus, the communication between the mobile terminal and the base station is shielded from the origin server. Split connection is employed by, for example, Indirect TCP (Bakre and Badrinath, 1997). TCP cwnd clamping (Chakravorty *et al.*, 2002) and WAP (WAP Forum, 1998) are examples of other technologies that use the split-connection concept.

End-to-end approaches only modify the communication endpoints. The philosophy is that the network should be kept as simple as possible. The purpose of the network is to transport packets, not to guess the requirements of the end-hosts. Therefore the end-hosts are where logic should be implemented. An example of an end-to-end approach is TCP Westwood (Casetti *et al.*, 2001), where a different algorithm is used to estimate the available bandwidth. The algorithm is more tolerant of the kinds of packet loss that occur in wireless networks, and hence improves performance. Further end-to-end proposals include the Eifel algorithm (Ludwig and Katz, 2000), WTCP (Sinha *et al.*, 2002), TCP Real (Zhang and Tsaoussidis, 2001), FreezeTCP (Goff *et. al.*, 2000), and delayed duplicate acknowledgements (Vaidya *et al.*, 1999).

10.6 CONCLUDING REMARKS

The future of multimedia for mobile devices looks bright. For example, more and more mobile phones come equipped with still-image cameras. The third generation mobile system makes video communication possible. Besides communication within the operator network, it is also desirable to interconnect to the Internet, to surf the web, send electronic mail, and so on.

Two fundamental ideas exist today for Internet access to wireless mobile devices. With loose coupling, the mobile network uses its own protocols and uses proxies for connection to the Internet. With tight coupling, the mobile devices get their own addresses, enabling IP communication, and can therefore interact with other hosts using end-to-end connectivity.

Full coupling may however pose performance problems. The Internet protocols were engineered with, for example, low non-congestion related packet losses and fairly stable round trip times in mind. As discussed in this chapter, wireless networks experience non-congestion related packet losses to a higher degree, and round-trip times may also vary a lot because of fluctuations in channel quality.

To this end, this chapter has investigated the possibility of increasing performance by examining the needs of multimedia applications versus the service that is provided by TCP. As multimedia applications may not always need the full reliability that TCP provides, this can be traded for better network performance. Experiments have been performed with an implementation of TCP-L, a modification of TCP to provide a more flexible transport service. The results of these experiments suggest that TCP-L can give better throughput than regular TCP, when residual bit errors are delivered from the network.

Some general open issues include the support of lower layers in delivering errors and cross-layer interaction. Since not many protocols make use of erroneous data from the link layer, the link layer has traditionally discarded such data. Further, with the advent of increased interaction between layers, the interface between layers needs to be studied further and defined.

10.7 REFERENCES

Alfredsson, S. (2001) 'TCP Lite—A Bit Error Transparent Modification of TCP', Master's thesis June, 2001, Karlstad University, Sweden.

Alfredsson, S. and A. Brunstrom (2003) 'TCP-L: Allowing Bit Errors in Wireless TCP', *Proceedings of IST Mobile and Wireless Communications Summit 2003*, pp. 149–154.

Allman, M., V. Paxson and W. Stevens (1999) *RFC 2581:* TCP Congestion Control.

Bakre, A. V. and B. Badrinath (1997) 'Implementation and Performance Evaluation of Indirect TCP', *IEEE Transactions on Computers*, pp. 260–278.

Balakrishnan, H., S. Seshan, E. Amir and R. H. Katz (1995) 'Improving TCP/IP Performance over Wireless Networks', *Proceedings of 1st ACM International Conference on Mobile Computing and Networking (Mobicom)*, pp. 2–11.

Balan, R. K., B. P. Lee, K. R. R. Kumar, L. Jacob, W. K. G. Seah and A. L. Ananda (2002), 'TCP HACK: A Mechanism to Improve Performance over Lossy Links', *Computer Networks,* **39**(4), 347–361.

Biaz, S. and N. Vaidya (1997) 'Using End-to-end Statistics to Distinguish Congestion and Corruption Losses: A Negative Result', *Technical Report 97-009*, Dept. of Computer Science Texas A&M University.

Casetti, C., M. Gerla, S. Mascolo, M. Y. Sanadidi and R. Wang (2001) 'TCP Westwood: Bandwidth Estimation for Enhanced Transport over Wireless Links', *Proceedings of ACM Mobicom*, Rome, Italy, pp. 287–297.

Chakravorty, R., J. Cartwright and I. Pratt (2002) 'Practical Experience with TCP over GPRS', *IEEE GLOBECOM*, pp. 1678–1682.

Ding, W. and A. Jamalipour (2001) 'A New Explicit Loss Notification with Acknowledgement for Wireless TCP', *Proceedings of PIMRC 2001*, Vol. 1, B65–B69.

Garcia, J. and A. Brunstrom (2002) 'Checksum-based Loss Differentiation', *Proceedings 4th IEEE Conference on Mobile and Wireless Communications Networks (MWCN 2002)*, Stockholm, Sweden.

Goff, T., J. Moronski, D. Phatak and V. V. Gupta (2000) 'Freeze-TCP: A True End-to-end Enhancement Mechanism for Mobile Environments', *Proceedings of IEEE INFOCOM*, Vol. 3, pp. 1537–1545.

Han, R. and D. Messerschmitt (1999) 'A Progressively Reliable Transport Protocol for Interactive Wireless Multimedia', *Multimedia Systems, 7*(2), 141–156.

Larzon, L.-Å., M. Degermark and S. Pink (1999) 'UDP Lite for Real-time Multimedia Applications', *Proceedings of the QoS mini-conference at the IEEE International Conference of Communications (ICC)*.

Ludwig, R. and R. H. Katz (2000) 'The Eifel Algorithm: Making TCP Robust Against Spurious Retransmissions', *ACM Computer Communication Review, 30*(1), 30–37.

Muratore, F. (2000) *UMTS—Mobile Communications for the Future*, John Wiley & Sons, Inc., New York.

Ostermann, S. (2003) *Tcptrace*, <http: //www.tcptrace.org/>

Parsa, C. and J. Garcia-Luna-Aceves (2000) 'Improving TCP Performance over Wireless Networks at the Link Layer', *ACM Mobile Networks and Applications Journal, 5*(1), 57–71.

Postel, J. (1980) *RFC 768*: User Datagram Protocol.

Postel, J. (1981a) *RFC 791*: Internet Protocol.

Postel, J. (1981b) *RFC 793*: Transmission Control Protocol.

Postel, J. (1981c) *RFC 801*: NCP/TCP Transition Plan.

Ratnam, K. and I. Matta (1998) 'WTCP: An Efficient Transmission Control Protocol for Networks with Wireless Links', *Proceedings*

Third IEEE Symposium on Computers and Communications (ISCC '98), Athens, Greece, pp. 74–78.

Rizzo, L. (1997) 'Dummynet: A Simple Approach to the Evaluation of Network Protocols', *ACM Computer Communication Review,* **27**(1), 31–41.

Sinha, P., T. Nandagopal, N. Venkitaraman, R. Sivakumar and V. Bharghavan (2002) 'WTCP: A Reliable Transport Protocol for Wireless Wide-area Networks', *Wireless Networks,* **8**(2), 301–316.

Stone, J. and C. Partridge (2000) 'When the CRC and TCP Checksum Disagree', *Proceedings of ACM SIGCOMM,* Stockholm, Sweden, pp. 309–319.

Vaidya, N., M. Mehta, C. Perkins and G. Montenegro (1999) 'Delayed Duplicate Acknowledgements: A TCP-unaware Approach to Improve Performance of TCP over Wireless', *Technical Report,* Computer Science Dept., Texas A&M University.

WAP Forum (1998) WAP Specifications for WAP version 1.0. <http://www.wapforum.org>

Zhang, C. and V. Tsaoussidis (2001) 'TCP Real: Improving Real-time Capabilities of TCP over Heterogeneous Networks', *Proceedings of the 11th IEEE/ACM NOSSDAV,* pp. 189–198.

11

Transcoding of Image Data

Johan Garcia and Anna Brunstrom

Department of Computer Science, Karlstad University

11.1 INTRODUCTION

As this book illustrates, the term 'multimedia' represents many different concepts. Undeniably it requires presentation of, and interaction with, media data of different formats. Media as consumed by humans is by its nature analog, since human perception ultimately is an analog process. However, the processing and transportation of media can be done with the media data in a digital format. The advent of digital media data and computers opened up many new possibilities with regards to creating, handling and distributing multimedia. Thanks to the losslessness of digital processing it is possible for digital media items to be replicated indefinitely without any media degradation, making simple distribution to a large amount of receivers possible. Distribution of digital multimedia data is nowadays, to a large extent, done over the Internet. As mentioned in the section on QoS in Chapter 8, every network has its

Perspectives on Multimedia R. Burnett, Anna Brunstrom and Anders G. Nilsson
© 2004 John Wiley & Sons, Ltd ISBN: 0-470-86863-5

limitations, and this is certainly true for the Internet. The basic problem is that the Internet is composed of a large number of networks with widely differing characteristics. Some of these characteristics, such as low bandwidth and large jitter, are undesirable when transporting real-time multimedia data. Attached to the networks is a large number of end hosts. The hosts can also have widely varying characteristics, as they range from supercomputers to hand-held computers and cellular phones. The ability to work despite the heterogeneity of both networks and end hosts is one of the major factors behind the success of the Internet, but this also makes it hard to establish one unified architecture for media distribution.

In order better to adapt media data to the conditions of a specific communications network or end device, it is possible to perform transcoding. Transcoding basically means recoding of a media object from one representation to another representation and is, as mentioned in Chapter 1, a consequence of the computerization of media. The media object can be of any type, and the transcoding of video, audio and images has been described in the literature. To provide a taste of the issues surrounding transcoding, this chapter discusses image transcoding in the web environment. From the user perspective, web browsing essentially entails the use of a web browser to retrieve and render web pages consisting of various web resources. The resources are typically HTML text or images, but other resources such as animation and video are not uncommon. In order to reduce the amount of data needing to be transferred, images are compressed using image compression. GIF (GIF, 1989) and JPEG (ITU-T, 1992) are the two most common standards for compression of web images. The compression can reduce the amount of bytes needed to represent an image to about 30–70 % (GIF) or 2–40 % (JPEG) of the original amount. Of the two standards, JPEG is the most complex, but also the most flexible. JPEG, for example, provides fine control over the compression level and the resulting image quality.

JPEG images used as web images tend to use moderate compression levels in order to provide excellent image quality. However, when using a slow-access network, many users would prefer to receive web images that are highly compressed. When receiving highly compressed images the user waiting time is considerably less than the waiting time for the moderately compressed images stored in the web server. Since the images stored in the web server have a moderate compression level, transcoding to the higher compression level must be performed. In addition to transcoding performed within one compression standard, as exemplified by the above transcoding of JPEG to JPEG with higher compression ratio, it is also possible to transcode from one standard to another. Transcoding can therefore be used to perform trade offs such as image quality versus download time, and also to adapt the media data to network- (or client-) specific characteristics such as high link-error ratios as detailed later in this chapter.

The transcoding is normally performed somewhere inside the network, between the original web server and the last hop to the user. Depending on the purpose of the transcoding, the best location for the transcoding varies. Figure 11.1 illustrates a case where the transcoding occurs at the boundary between the wired Internet and a wireless access network. A transcoding proxy is used to host the transcoding functionality. A number of such proxies has been discussed in the literature, having various design and functionality, for example Mowgli (Kojo *et al.*, 1994), Pythia (Fox and Brewer, 1996) and MOWSER (Bharadvaj *et al.*, 1998).

The remainder of this chapter begins with a short overview of JPEG coding and then presents various efficient techniques for compression level transcoding of JPEG images. After that comes a section describing transcoding from JPEG to robust JPEG, which is a JPEG variant designed to handle packet loss better. Towards the end of this chapter, further reading is suggested, followed by some concluding remarks.

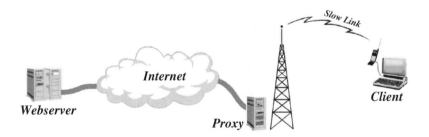

Figure 11.1 Example proxy setup

11.2   JPEG CODING

JPEG (ITU-T, 1992) is a relatively complex standard that can be operated in a number of modes. Our interest here is primarily aimed at transfer of web-type images, and these are typically coded using either sequential baseline mode or progressive mode. Sequential encoding provides the user with an image that grows from top to bottom as more data arrive. Progressive encoding, instead, quickly provides the user with a coarse image that is step-wise refined as more data are received. This is illustrated in Figure 11.2 which shows how sequentially and progressively coded images look after receiving 30 % of the total image data. When all data have been received, both the sequentially and progressively coded image will look like the image in Figure 11.3.

In order to provide some background for the later presentation of transcoding techniques, a short review of sequential baseline JPEG is first provided to explain the basic operation of JPEG. The encoding of an image comprises the following steps:

(1) Make a colour conversion that changes the image representation to the YC_bC_r colour space. Images on the screen are represented by the RGB colour space, but the YC_bC_r colour space is better suited for later compression. The YC_bC_r colour space consists of three components, the luminance (or brightness, Y) component and two chrominance (or colour, C_b and C_r) components.

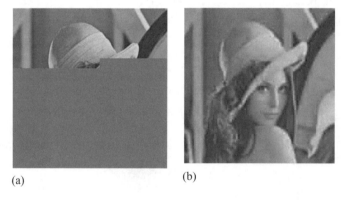

(a)   (b)

Figure 11.2 Image after receiving 30 % of data: (a) sequential coding; (b) Progressive coding

Figure 11.3 Image after receiving 100 % of data

(2) Downsample the colour components. The human visual system is less sensitive to colour, and a considerable reduction in data size can be achieved by exploiting this. A typical downsampling is 2×2 which replaces a 2×2 square of pixels with one pixel that holds the mean value of the four original pixels. After this downsampling the two chrominance components will have a quarter as many pixels[†] as the luminance component.

(3) Split the components into 8×8 pixel blocks and make a discrete cosine transformation (DCT) on each block. Instead of 64 pixels, the DCT creates 64 coefficients which are divided into one DC coefficient holding the average value of the pixels in the block and 63 AC coefficients holding the amount of different spatial frequency patterns. Figure 11.4 shows the appearance of the 64 different spatial frequency patterns. Somewhat simplified, a coefficient specifies 'how much' the corresponding DCT frequency pattern is present in the original pixel block.

(4) Quantize the coefficients according to the specified quantization tables. Each entry in the quantization table controls the level of precision used to represent the corresponding coefficient value. The limit of human perception is different for the different spatial frequencies in images. The quantization tables reflect this fact by doing more aggressive quantizing of coefficient values corresponding to patterns with high spatial frequencies.

[†] The term used in the standard is sample, but this simplified presentation will use the term pixel.

DC Coefficient

Increasing horizontal frequency

Increasing vertical frequency

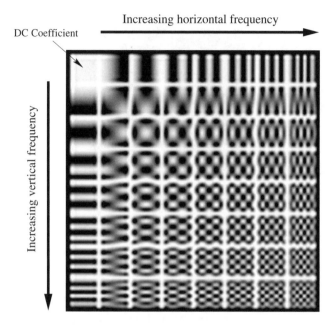

Figure 11.4 DCT spatial frequency patterns

(5) Code the DC coefficient using predictive coding with the DC value of the previous block as the predictor. Run-length encode the quantized coefficients, grouped together in a zigzag order to ensure longest possible zero run lengths.

(6) Huffman encode (Huffman, 1952) the run-length encoded coefficients. The encoding is done either according to a table suggested in the standard, or by making a pass through the data, which produces a table tailored to the specific image. This latter approach is commonly called Huffman table optimization and provides an increase in compression at the expense of the additional computation needed to collect the statistics.

(7) Packetize the data according to an interchange format. JFIF (Hamilton, 1992) is the most common format.

The baseline JPEG described above is a lossy coding, which means that the image produced after encode and decode processing is not exactly identical to the original image. As evident from the description of the coding, it is designed to lose information where the human perception is least sensitive to loss. Information loss occurs in steps (2) and (4) only. In step (2), the amount of downsampling partly decides the amount of information lost in the chrominance components. In step (4), a quantization table decides the coarseness used in representing each coefficient. Each of the 64 DCT coefficients has its own quantization value in the table. The quantization value decides the granularity used to represent the coefficient. In this step it is possible to trade a reduction in image data size against reduced image quality by using larger values when quantizing. This quantization table scaling can be indirectly controlled with the quality setting in most image manipulation programs. A further reduction of the data size occurs in steps (5) and (6), but this is achieved by entropy coding algorithms that do not lose information.

Step (7) formats the encoded image data, typically according to JFIF, which places additional header information in the beginning of the file. The JFIF header is followed by the quantization and Huffman tables necessary to perform decoding. After the header data follows the Huffman coded data representing the image. The exact data organization is dependent on the downsampling used. A typical configuration is 2×2 downsampling, which causes the data to be organized as four Y blocks, then one C_b block and one C_r block. In this case there will be six blocks grouped together. This group of associated blocks is called a minimum coded unit (MCU) and is used as an atomic unit in the output data stream. In this example, the data in one MCU represents a 16×16 pixel area in the output image.

The steps described above provide an insight into how sequential JPEG encoding is performed. The decoding is, simply put, performed by doing the inverse of each step proceeding backwards from step (7) to step (1). As stated at the beginning of this section, the above description applies to sequential images. A progressive image is different in that it is composed of a number of scans. Each scan provides one step of refinement to the image. The progressive mode of JPEG allows two mechanisms for obtaining the progressiveness, namely spectral selection and successive approximation. Spectral selection is performed by sending only a subset of the DCT coefficients in a scan. Successive approximation sends only a few of the most significant bits in one scan, sending more bits in subsequent scans. The above mechanisms can be combined, and typical progressive images contain such combined progression sequences. Progressive images can be seen as performing the steps run-length coding, Huffman coding and packetization in several iterations, one iteration per scan.

11.3   JPEG COMPRESSION TRANSCODING

As discussed in the Introduction, transcoding can be used to reduce user waiting times. Before the image data traverses a slow link, transcoding from JPEG to JPEG with higher compression is one way to reduce the user waiting times. This transcoding is done in a transcoding proxy. However, the transcoding of images requires considerable processing resources in the transcoding proxy. It is therefore important to perform the transcoding as resource-efficiently as possible. This section presents two different approaches for efficient JPEG compression level transcoding: one that applies to sequentially encoded images, and one that applies to progressively encoded images. Both of these approaches are based on using knowledge about the JPEG encoding process to, in effect, short-circuit some of the more complex JPEG encoding steps, thus creating efficient transcoding.

11.3.1   Sequential JPEG

When doing regular transcoding a full decoding cycle is followed by a full encoding cycle, as illustrated by the thin solid line in Figure 11.5. Each box in the figure corresponds to one of the steps described in the previous section. Regular transcoding uses a pixel-based format as the middle format between decoding and encoding. A pixel-based format must be used when two different compression techniques are used, for example, when transcoding between JPEG and some different compression standard. However, when transcoding between different compression levels within the JPEG compression technique, the use of a pixel-based middle format is no longer mandatory. For sequential JPEG images, a more suitable middle format is the dequantized DCT coefficients. By doing the transcoding using the DCT coefficients as middle format, a DCT domain transcoder achieves a considerable reduction in processing requirements for the transcoding of any given JPEG image.

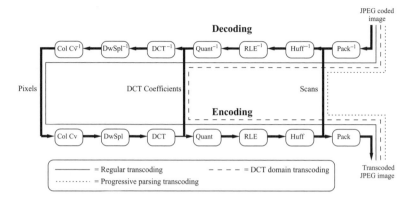

Figure 11.5 JPEG transcoding approaches

The DCT domain transcoder exploits the fact that the three leftmost decoding steps in Figure 11.5 are necessary only if a pixel-based image is to be produced. With the exception of chrominance downsampling, the output compression level is decided in step (4), which makes it possible to use the DCT coefficients of the image present between steps (3) and (4) as the middle format. As illustrated by the dashed line in Figure 11.5, this allows the DCT domain transcoder to bypass the three left-most decoding and encoding steps, saving a considerable amount of processing time primarily by not performing the DCT. If the original image is not downsampled, and downsampling is desirable for the output image, the downsampling of the chrominance components can be effectively done in the DCT domain as demonstrated by Dugad and Ahuja (2001). Further details on DCT domain processing in general are available in Smith and Rowe (1996) and de Queiroz (1997).

To provide some insights into the effectiveness of DCT domain transcoding with regards to processing requirements, three different transcoders have been experimentally evaluated.

- A modified version of the jpegtran program. This program is part of the IJG jpeglib 6b distribution (*Independent JPEG Group Software*, n.d.) and was originally developed to do lossless image transformations such as rotation. The program was extended to also do DCT domain compression level transcoding as described earlier.

Table 11.1 Image test sets

Test set	No. of images	Image size	Data amount
1	20	30×30	16.4 kb
2	1	718×1015	233 kb
3	8	154×196–507×800	221 kb

- The djpeg/cjpeg (v. 6b) programs which are part of the same IJG jpeglib distribution as the jpegtran program. Compression level transcoding was accomplished by piping the output of the decoder (djpeg) to the input of the encoder (cjpeg).

- The convert program (v. 4.1.0), which is a generic image conversion program capable of transcoding between many compression formats and of transcoding between different JPEG compression levels.

The three test sets shown in Table 11.1 were used. One test set was chosen to reflect typical web images (set 3), and two sets of atypical images were used in order to test boundary behaviour (sets 1 and 2).

All runs were performed on the same computer running Linux. Twenty-five runs were performed for each test. The mean values for the tests are shown in Figure 11.6. The 99 % confidence intervals are too small (0.02) to be shown in the figure.

The measured times include time for process creation, file I/O and other relatively fixed overheads. The experimental data supports the intuition that this overhead is important when the file sizes are small, as in test set 1. When using larger files, as in test set 2, the gain provided by efficient transcoding becomes evident. Test set 3 is meant to reflect the usage that could occur in a proxy, with a mix of images of different sizes. For test set 3, the DCT domain recoder is considerably more efficient than the djpeg/cjpeg approach, which requires over 90 % more processing time. The performance improvement of the DCT domain approach is evident considering that the DCT domain recoder shares the same code base as djpeg/cjpeg. The convert recoder performs even worse and needs over 200 % more processing time.

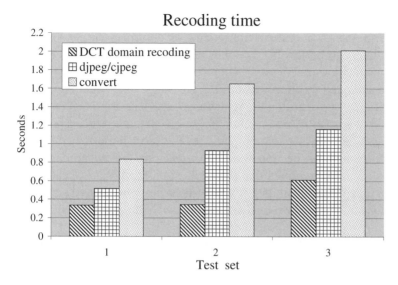

Figure 11.6 Transcoding time measurements

In addition to the measurements on processing requirements shown above, measurements on the memory requirements have been performed. These measurements show that the more general transcoder, convert, requires considerably more memory than either cjpeg/djpeg or DCT domain transcoding. Further details on the memory measurements are available in Garcia and Brunstrom (2000a). In conclusion, the performance results indicate that the DCT domain recoder can obtain a transcoding throughput that is considerably higher than alternative approaches while still having low memory requirements.

11.3.2 Progressive JPEG

From a transcoding point of view, the biggest difference between progressive and sequential JPEG images is in the different data organization. For sequential images the received data builds up the image row by row from top to bottom. Progressive images are instead composed of a number of scans, where each scan contains a step-wise refinement of the whole image. This difference has implications for compression level transcoding. For sequential input images, transcoding can be performed either by doing a full decoding/encoding cycle or, more efficiently, by performing DCT domain transcoding as described earlier. Both these methods can be labeled as being requantization based. In addition to requantization-based transcoding, progressive input images can also be transcoded by simply truncating the image data stream at a suitable position and discarding the remaining data. This approach exploits the inherent scalability potential of progressive images. Since a progressive data stream always contains the visually most important information first, images can be transcoded by truncating the image data stream. This of course requires that the application receiving truncated image data is capable of handling it. Our tests have shown that all tested applications, including two major browsers, have the capability of correctly displaying truncated progressive images.

This method of transcoding progressive JPEG images is called progressive parsing transcoding, and we provide a more detailed description in Garcia and Brunstrom (2000b). Progressive parsing transcoding is simpler and faster than other suggested transcoding approaches and can also provide a better rate/distortion performance. The reasons why progressive parsing may yield transcoded images with a better quality for a given compression ratio is discussed in Garcia and Brunstrom (2000b) and is also briefly presented below.

11.3.2.1 Quality improvements   Figure 11.7 shows a section of the well-known Lena image. Table 11.2 shows the details for these images, providing the compression ratio in bits per pixel (bpp). The PSNR (peak signal-to-noise ratio) of the transcoded images relative to the original progressive image is also shown as one metric for image quality. Higher PSNR is better.

As the PSNR values suggest, Figure 11.7 shows that the progressively parsed image has higher visual quality than the sequentially transcoded image, this being especially visible in the chin area. The better quality is even more noticeable in colour reproduction. The quality improvement comes from three factors, one applicable to progressively coded images in general and two specific for progressively parsed images.

One factor is the Huffman table optimization, which is always performed for progressive images. For sequential images, the Huffman

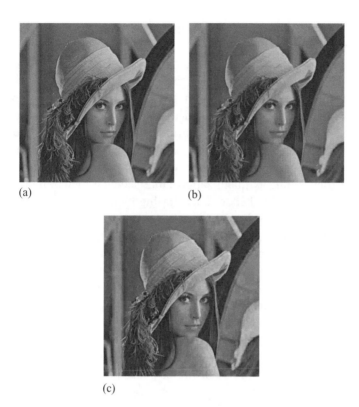

(a)   (b)

(c)

Figure 11.7 Quality illustration: (a) original progressive; (b) progressively parsed; (c) sequential transcoded

Table 11.2   Image compression level and quality

Image	bpp	PSNR	IJG quality setting
Original progressive	0.835	—	60
Progressively parsed	0.281	31.402	60 + skewing
Sequential transcoded	0.296	30.677	10

tables proposed in the JPEG standard are typically used since this saves processing and memory resources by not performing Huffman table optimization during the encoding. The increased compression performance provided by Huffman table optimization is obviously also present in progressively parsed images. When performing progressive parsing transcoding, this performance advantage does not incur any extra resource usage at the proxy. The Huffman table optimization done when the original image was first progressively encoded is in effect reused.

Another factor contributing to the quality improvement in progressive parsing transcoding is quantization table skewing. This effect makes the quantization table better suited for highly compressed images and can provide a quality improvement relative to images transcoded by requantization-based methods. This effect is a result of the fact that the relative importance of the individual coefficient quantization values changes as scans arrive.

Another advantage of using progressive parsing transcoding, as opposed to requantization-based transcoding, is that less noise is injected by quantization value changes. This effect occurs because the effective quantization values for progressive parsing are always multiples of the original quantization values, whereas requantization results in changes that are not whole multiples of the original quantization values. This means that the zero-error-accumulation property (Chang and Eleftheriadis, 1994) holds for more coefficients, thus inducing less total requantization noise when performing progressive parsing.

11.3.2.2 Scan selection

11.3.2.2 Scan selection   Progressive parsing transcoding provides advantages from both resource usage, delay and image quality perspectives. The question remains, however, of how to best implement the transcoding in practice. Progressive parsing is centered around the truncation of an incoming data stream in order to achieve a target compression level, C_g (bpp). The number of pixels in an image is easily computed using the image width, W, and height, H, that are given in the JPEG header. This information is then used to perform the truncation in different ways; simple truncation, inter-scan truncation or intra-scan truncation.

Simple truncation is the most basic form of progressive parsing transcoding. An end-of-image marker (EOI) is simply inserted into the data stream when C_g is reached, i.e. after WHC_g bits. The remaining data are discarded. Although simple, this method has a drawback because it will not provide a consistent quality level over the whole image. When the data stream is truncated in the middle of a scan, the upper part of the picture will have slightly better quality than the lower part. Depending on the scan granularity used, this effect will be more or less visible.

With inter-scan truncation the data stream is instead truncated after n of the k total scans by inserting an EOI marker after the nth scan and discarding the $k - n$ last scans. The output compression level is thus only variable in k discrete levels. Inter-scan truncation provides the same quality level over the whole image, but the resulting compression level will not be exactly C_g.

Intra-scan truncation is a refinement of inter-scan truncation that, instead of truncating at a scan boundary, truncates inside a scan in a way that upholds the progressiveness. By buffering each scan and detecting the scan n in which C_g is reached, scan n can then be transcoded. By performing Huffman and run-length decoding on this scan, it becomes possible to trim the scan so that the resulting compression level is a close match of C_g.

To summarize, truncation, as done in progressive parsing transcoding, leads to very efficient transcoding that also shows some image quality benefits over other approaches.

11.4 ROBUST JPEG TRANSCODING

Whereas the previous section describes how best to perform JPEG transcoding in order to increase the compression ratio, this section provides one example of transcoding that makes the data stream more robust to data loss. This section presents a transcoder that outputs a data stream that is based on JPEG but is modified to provide better results when experiencing data losses during transfer. Although robust JPEG is based on standard JPEG, it employs mechanisms, such as interleaving, which require a robust JPEG decoder to successfully decode the data stream. This is in contrast with output from the transcoders discussed in the previous section, which are decodable by standard JPEG decoders.

There exist many ways of modifying JPEG to make it more robust. The design of the described coder is based on the assumption that it will use the partially reliable transport service provided by the PRTP protocol described in Chapter 9. The intended application area is web browsing, and the data-loss type considered is packet loss, not bit errors or bit erasures. The coder must be able to handle the loss of one or more packets of variable size, and still resynchronize and conceal the loss as well as possible. The output quality from the coder should degrade gracefully as the loss rate increases. In order to adapt the JPEG coding for partial reliability a strategy consisting of three steps is employed: (1) Extend the resynchronization capabilities of regular JPEG, (2) Perform interleaving so that the lost data are not aggregated in one place in the image, (3) Try to conceal the lost data by using redundancies not removed by the source coding.

The steps are further explained below, and their effects are illustrated in Figures 11.8–11.12. The images in the figures were produced by the robust coder and used an original web image of typical quality as input. The original image is shown in Figure 11.8, and Figure 11.9 shows the same image after three packet losses resulting a 10 % data loss.

11.4.1 Decoder resynchronization

Resynchronization is needed to allow the decoder to reach a known state after a data loss. For a byte-stream-oriented transport, such as PRTP, losses should be assumed to be of unknown length as the application has no control over the segmentation of data. The optional resynchronization capability present in JPEG is based on the periodical insertion of restart markers into the data stream. The markers reset the predictive coder used for the DC coefficient and allow the following data to be placed in the correct position in the image. However, there are only eight restart markers specified by the JPEG standard. The small number of unique markers necessitates a trade off. On one hand the restart markers need to be sufficiently spread out to ensure that a loss does not cause marker number wraparound, thus making unambiguous positioning impossible. On the other hand, the markers should be spaced closely to minimize the amount of wasted data, since data received after a loss but before a restart marker is unusable. To resolve this issue we introduce the notion of extended restart markers which, instead of eight unique markers, provides 190 markers. This allows unambiguous positioning for all practical image sizes while causing little wasted data. Since the extension of the restart

Figure 11.8 Original image

markers uses unused JPEG marker space, the extended restart markers provide enhanced resynchronization without having to use longer markers. The effect of using extended resynchronization markers versus no markers is illustrated in Figures 11.9 and 11.10.

11.4.2 Interleaving

Interleaving is performed with the objective of distributing the effects of packet losses as evenly as possible within the image in order to facilitate error concealment. In addition to distributing losses as evenly as possible, it is desirable to minimize the chance of separate packet losses causing several losses of neighbouring blocks in the image. Interleaving is often done by redistributing the blocks of the image, but it is also

Figure 11.9 Original image after loss

possible, instead or additionally, to redistribute the coefficients in the blocks (Chang, 1998). There exist several block interleaving strategies, and one strategy is to try to spread the losses evenly, based on network and image characteristics. The maximum spreading interleaver, described in more detail in Garcia (2002), uses the length of a typical data loss (i.e. packet size), the pixel width and height of the image, the compression ratio and the amount of downsampling used for different components. Using these values, and compensating for the number of MCUs between restart makers, the maximum spreading interleaver calculates an interblock distance, which ensures that a packet loss is spread evenly over the whole image. However, if a second packet loss occurs so that the deinterleaved blocks neighbour previously lost blocks, then many blocks which were lost in the first packet loss may lose

Figure 11.10 Extended restart

neighbouring blocks. This occurs because both losses are interleaved with a fixed interblock distance. Such repeated absence of neighbouring blocks is undesirable since it hampers the performance of most error concealment algorithms.

In order to minimize the probability of multiple data losses incurring a stride of lost neighbouring blocks, it is possible to use random block interleaving. This interleaving uses a random number generator to obtain the interleaved position of each block. By distributing the blocks randomly, the long strides of lost blocks that could occur using maximum spreading are removed. However, it is possible to create spurious neighbour block losses even for single data losses. Random block interleaving is illustrated in Figure 11.11.

Figure 11.11 Random block interleaved

In order to improve the random block interleaving, we have devised a window-based random interleaving, which is capable of alleviating the neighbour block loss problem for single data losses. The detailed algorithm is available in Garcia (2002).

11.4.3   Error concealment

Error concealment tries to hide lost information by using information from surrounding blocks. As shown earlier, JPEG uses DC and AC coefficients to code each 8×8 block.

The DC coefficient contains the average colour of the block. If the DC coefficient differs too much from the correct value, the edges around the block may become noticeable. These edges are important to the

Figure 11.12 Random block + concealed

perceived image quality since they are straight line artefacts, which are not well masked by the visual system. We base our reconstruction on a method mentioned in Ridge *et al.* (1999), which compares the difference between the two vertical neighbour blocks with the difference between the two horizontal neighbour blocks. The two blocks that have the lowest difference are averaged to obtain the interpolated value.

The AC coefficients contain information on the presence of different spatial frequency contents in the block. This fact is used when performing the interpolation as described in Chang (1998). The AC coefficients can be divided in three categories: (i) containing mainly horizontal frequency information, (ii) containing mainly vertical frequency information, or (iii) containing both. Accordingly, a coefficient is interpolated from different surrounding blocks depending on the category the coefficient belongs to. The effect of error concealment is visible in Figure 11.12.

11.5   FURTHER READING

In this chapter, transcoding is exemplified by JPEG transcoding. JPEG is a complex standard and, due to space limitations, the overview of JPEG is kept very brief in this text. To provide further understanding, Pennebaker and Mitchell (1993) provide a very complete description of JPEG. A shorter presentation is provided by Wallace (1991). Image transcoding can be used for many applications and the web is one typical application. The usefulness of transcoding for web images is discussed by Chandra *et al.* (2000). As mentioned in the introduction, transcoding can be applied to many multimedia data types in addition to the image transcoding discussed in this chapter. Video transcoding has been researched for quite some time and an example of early work on video transcoding is that of Amir *et al.* (1995). Vetro *et al.* (2003) provide a nice overview of current video transcoding issues. A more general discussion on media adaptation is provided by Bolliger (2000) and also by Margaritidis and Polyzos (2001).

11.6   CONCLUDING REMARKS

Not only is the use of multimedia spreading to new types of devices, but the distribution of multimedia data is also done using a range of new networking technologies. This creates an increasingly heterogeneous multimedia environment. Transcoding is one way of addressing the increased heterogeneity and tailoring for the differences in presentation devices and distribution network technologies. By applying efficient transcoding it is possible to influence trade offs such as that between image quality and download time. It is also possible to use transcoding to adapt to specific network characteristics such as high packet or bit error rates.

This chapter illustrates how knowledge about the JPEG coding process can be used to create a transcoder that requires only 30 % of the processing required for a general transcoder. Transcoding in the case of transport services not providing reliable transport of the compressed image data is also discussed. These examples illustrate how transcoding can be used to adapt one type of media content. However, this is a general concept and transcoding of different types of media can help to overcome the challenges of an increasingly heterogeneous environment.

11.7 REFERENCES

Amir, E., S. McCanne and H. Zhang (1995) 'An Application Level Video Gateway', in *Proceedings of ACM Multimedia*, San Francisco, CA, pp. 255–265.

Bharadvaj, H., A. Joshi and S. Auephanwiriyakul (1998) 'An Active Transcoding Proxy to Support Mobile Web Access', *Proceedings of the 17th IEEE Symposium on Reliable Distributed Systems*, pp. 118–123.

Bolliger, J. (2000) *A Framework for Network-aware Applications*, PhD thesis, ETH Zürich.

Chandra, S., C. S. Ellis and A. Vahdat (2000) 'Differentiated Multimedia Web Services Using Quality Aware Transcoding', *Proceedings of INFOCOM*, Tel-Aviv, Israel, pp. 961–969.

Chang, E. Y. (1998) 'An Image Coding and Reconstruction Scheme for Mobile Computing', in T. Plagemann and V. Goebel (eds) *Proceedings IDMS*, Oslo, Norway, pp. 137–148.

Chang, S.-F. and A. Eleftheriadis (1994) 'Error Accumulation of Repetitive Image Coding', *Proceedings of IEEE International Symposium on Circuits and Systems*, pp. 201–204.

Dugad, R., and N. Ahuja (2001) 'A Fast Scheme for Image Size Change in the Compressed Domain', *IEEE Trans. on Circuits and Systems for Video Technology*, **11**(4), 461–474.

Fox, A. and E. A. Brewer (1996) 'Reducing WWW Latency and Bandwidth Requirements by real-time Distillation', *Computer Networks and ISDN Systems*, **28**(7-11), 1445–1456.

Garcia, J. (2002) *Application and Transport Layer Flexibility: An Image Transfer Example*, Karlstad University Studies 2002:11, Karlstad, Sweden.

Garcia, J. and A. Brunstrom (2000a) 'Efficient Image Transfer for Wireless Networks', *Proceedings of the 2nd International Conference on Advanced Communication Technology (ICACT2000)*, Muju, South Korea, pp. 305–310.

Garcia, J. and A. Brunstrom (2000b) 'Progressive Parsing Transcoding of JPEG Images', *Proceedings of 7th International Workshop on Mobile Multimedia Communications (MoMuC2000)*, Tokyo, Japan, pp. P-23-1–P-23-5.

Gif (1989) Graphics Interchange Format, version 89a, *Technical Report*, Compuserve Incorporated, Columbus, Ohio.

Hamilton, E. (1992) *The JPEG file interchange format*, C-Cube Microsystems, Inc. <http://www.ijg.org/files/jfif.ps.gz>

Huffman, D. A. (1952) 'A Method for the Construction of Minimum Redundancy Codes', *Proceedings of the Institute of Electronics and Radio Engineers*, **40**, 1098–1101.

Independent JPEG Group Software (n.d.) <http: //www.ijg.org>

ITU-T (1992) *Recommendation T.81—digital compression and coding of continuous-tone still images*, Geneva, Switzerland.

Kojo, M., K. Raatikainen and T. Alanko (1994) *Connecting Mobile Workstations to the Internet over a Digital Cellular Telephone Network*, University of Helsinki, Department of Computer Science, Series of Publications C, No. C-1994-39.

Margaritidis, M. and G. C. Polyzos (2001) 'Adaptation Techniques for Ubiquitous Internet Multimedia', *Wireless Communications and Mobile Computing*, (1), 141–163.

Pennebaker, W. B. and J. L. Mitchell (1993) *JPEG Still Image Data Compression Standard*, Van Nostrand Reinhold, New York.

Queiroz, R. de (1997) 'Processing JPEG-compressed Images', *Proceedings of IEEE International Conference on Image Processing (ICIP)*, Vol. II, Santa Barbara, CA, pp. 334–338.

Ridge, J., F. W. Ware and J. D. Gibson (1999) 'Image Refinement for Lossy Channels with Relaxed Latency Constraints', *Proceedings of IEEE Wireless Communications and Networks Conference 1999 (WCNC'99),* Vol. 2, pp. 993–997.

Smith, B. C. and L. A. Rowe (1996) 'Compressed Domain Processing of JPEG-encoded Images', *Real-Time Imaging,* **2**(1), 3–17.

Vetro, A., C. Christopoulos, and H. Sun (2003) 'Video Transcoding Architectures and Techniques: An Overview', *IEEE Signal Processing Magazine,* **20**(2), 18–29.

Wallace, G. K. (1991) 'The JPEG Still Picture Compression Standard', *Communications of the ACM,* **34**(4), 30–44.

About the Authors

STEFAN ALFREDSSON

Stefan Alfredsson is a lecturer and PhD student in the Department of Computer Science at Karlstad University, Sweden. He received his MS degree in computer science from Karlstad University in 2001. His research interests include transport layer aspects in wireless networking and increasing cross layer communication.

KATARINA ASPLUND

Katarina Asplund is a PhD candidate and lecturer at the Department of Computer Science at Karlstad University, Sweden. She received her MS degree in computer science from Karlstad University in 2000. Her area of interest is computer networking, especially transport layer support for multimedia communication and user perception of quality-of-service.

Perspectives on Multimedia R. Burnett, Anna Brunstrom and Anders G. Nilsson
© 2004 John Wiley & Sons, Ltd ISBN: 0-470-86863-5

LARS ERIK AXELSSON

Lars Erik Axelsson is a lecturer and a doctoral candidate in information systems. He is one of the originators of the multidisciplinary multimedia programme at Karlstad University. Lars Erik has also been the director of studies in the Department of Information Systems for more than a decade. His research interest is the analysis and design of information for data-base management, including theories and methods for processing information from various data sources.

ROBERT BURNETT

Robert Burnett is professor of Media and Communication Studies, Karlstad University. Burnett is the author of *The Global Jukebox:The International Music Industry* (Routledge, 1996), *Concentration and Diversity in the International Phonogram Industry* (Gothenburg, 1990) and the co-author of, *Web Theory* (Routledge, 2002, with David Marshall).

His main research areas include the global media, the music industry, digital media, and the Internet.

Burnett has been a regular commentator on global media issues with appearances on radio and television programs and commentaries for newspaper and magazine articles. He is one of the founders of the Communication, Media and Information Technology (CMIT) Research Group as well as the HumanIT Research Group. Burnett has been a visiting scholar in Canada, the USA, Scotland, Norway and Zimbabwe.

ANNA BRUNSTROM

Anna Brunstrom received MSc and PhD degrees in computer science from College of William and Mary, Virginia, USA in 1993 and 1996, respectively. In 1996 she joined the Department of Computer Science at Karlstad University where she is now a professor. She is research manager for the distributed systems and communications research group and also served as research manager for the Department of Computer Science from July 1998 until June 2002. She leads several research projects on communication protocols and is active in national and international projects on wireless communications. She has authored and co-authored two book chapters and over 30 scientific papers. She is a member of IEEE and ACM.

JOHAN GARCIA

Johan Garcia received a MS degree and the licentiate degree in computer science from Karlstad University in 2000 and 2002, respectively. He is currently a lecturer and PhD candidate in the Department of Computer Science at Karlstad University. His areas of interest are in networking, with special emphasis on wireless and multimedia communications, cross-layer issues and flexible transport protocols. He is a member of IEEE and ACM.

STEVE GIBSON

Steve Gibson is a Canadian multimedia artist, composer, and theorist. He currently is associate professor in digital media at the University of Victoria, Canada and is conducting research on real-time networked performance. Simultaneously deeply involved with technology and deeply suspicious of it's effects, Gibson's musical, multimedia and virtual reality work celebrates both the liberation and paranoia of techno-fetishism. His installations and compositions have been performed in such venues as Ars Electronica, the Whitney Museum of American Art, the North American New Music Festival, the European Media Arts Festival, ISEA 95, and 4 and 6CyberConf. His work has been published internationally by St Martin's Press, MIT Press, New World Perspectives, Turnaround Productions, Future Publications, Urra Apogeo, and Passagen Verlag.

ERLAND JONSSON

Erland Jonsson is professor of computer security and past head of the department of computer engineering at Chalmers University of Technology. Prior to taking up his present post, he worked in industry for almost 20 years on hardware and software design and quality assurance for telecommunications and space applications. His research interests include issues regarding the quantitative assessment of security, analysis of systems with simultaneous requirements on security, reliability and safety as well as intrusion detection systems.

ANDREAS KITZMANN

Andreas Kitzmann is currently a senior lecturer in the Department of Communication Studies at Karlstad University, Sweden. He received his PhD in comparative literature from McGill University where he did his dissertation work on the literary uses of hypertext. His research interests include the impact of communications technology on the construction and practice of identity, electronic communities, and the influence of new media on narrative conventions. Among his publications is the book *Saved From Oblivion: the Place of Media, from Diaries to Web Cams* (Peter Lang, forthcoming), which concerns the autobiographical uses of media technology ranging from hand written diaries to web-cams.

STEFAN LINDSKOG

Stefan Lindskog is a lecturer at the department of computer science at Karlstad University in Sweden. He has a licentiate of engineering degree in computer engineering from Chalmers University of Technology in Göteborg. His research focuses on computer and network security, especially methods for analysis and categorization of intrusions and vulnerabilities. He is also co-author of a textbook on web site privacy in P3P. Stefan is a member of IEEE.

LENNART MOLIN

Lennart Molin is a lecturer and doctoral student in information systems at Karlstad University, Sweden. He has a licentiate degree in information systems and his research concerns requirements specification for multimedia systems. In his current main project, one research question concerns how an experimental prototyping tool can be useful in transforming unclear wishes to system-specific requirements. A related topic is how to support end user involvement in this phase of multimedia development.

ANDERS G. NILSSON

Anders G. Nilsson is professor of information systems at Karlstad University, Sweden. He has a PhD in information management from Stockholm School of Economics, and is a research partner at the Institute for Business Process Development (Institute V) in Stockholm. Anders has been working as a researcher with the ISAC approach to information systems development, the SIV method for acquisition of standard application packages (enterprise systems, ERP systems) and a business modelling framework for creating method combinations. He has also been active as an adviser to many change projects in private industry and the public sector, and has been for many years deputy chair for the special interest group for systems development (SIG-SYS) at the Swedish Information Processing Society. Anders is author/co-author of 14 books on business and systems development.

JOHN SÖREN PETTERSSON

John Sören Pettersson is associate professor in information systems and the director of the Centre for HumanIT at Karlstad University, Sweden. His research concerns requirements specification for interactive systems in a wide range of environments such as developing countries and special education, as well as within multimedia products. A specific topic is how to support end user involvement in systems development processes.

LOUISE ULFHAKE

Louise Ulfhake is a lecturer and PhD student in information systems at Karlstad University, Sweden. Her research interest is aimed at models and methods for developing and evaluating interactive multimedia programs as educational software. Her interests include distance education using modern IT support. She served as programme director of a multidisciplinary multimedia programme at Karlstad University in 1996–1997, and as chair of the Committee for Life Long Learning in 1997–2001.

Index